TRANSACTIONS

OF THE

AMERICAN PHILOSOPHICAL SOCIETY

HELD AT PHILADELPHIA

FOR PROMOTING USEFUL KNOWLEDGE

NEW SERIES—VOLUME XXIV

PART 1.   APRIL, 1934

A New Marsupial Saber-tooth from the Pliocene of Argentina and its
Relationships to Other South American Predacious Marsupials

Elmer S. Riggs

PHILADELPHIA:

THE AMERICAN PHILOSOPHICAL SOCIETY

104 South Fifth Street

1934

LANCASTER PRESS, INC., LANCASTER, PA.

A NEW MARSUPIAL SABER–TOOTH FROM THE PLIOCENE OF ARGENTINA AND ITS RELATIONSHIPS TO OTHER SOUTH AMERICAN PREDACIOUS MARSUPIALS

By ELMER S. RIGGS

WITH APPENDIX BY BRYAN PATTERSON

CONTENTS

LIST OF PLATES

TEXT FIGURES

INTRODUCTION

THE Marshall Field Paleontological Expedition, working in the Province of Catamarca, Argentina, in 1926, made extensive collections of fossil mammals from the Araucanian formation of that region. These collections include some 200 specimens, of which number eleven are of carnivores and carnivorous marsupials. The true carnivores are

1

limited to immigrant members of the Procyonidæ (*Amphinasua*), related to the recent coatis but of larger species. The didelphid marsupials are represented in the collections by a single small species, *Didelphis biforata* Ameghino. The common type of borhyænid marsupial * is recognized in a single species of intermediate size, *?Acrohyænodon acutidens* Rovereto. A second group of marsupial carnivores of large size and of characteristics unknown until the discovery of specimens below recorded, forms the chief subject of this paper. A preliminary report of this discovery † was made to the Paleontological Society of America in December, 1928, but no name was given to the group at that time. A brief description was later published by the Field Museum.‡

The specimens have been prepared and studied in comparison with other marsupials of the same formation as well as with the better-known genera of borhyænid marsupials from the Santa Cruz formation. Comparisons have also been made with *Thylacynus* and *Dasyurus* among living marsupials, and with *Hyænodon* and other creodonts. Obligations are here expressed to the American Museum and to the U. S. National Museum for loan of specimens used in study; also to the staff preparators and artists of Field Museum for their skillful services in preparing the specimens and the illustrations and to Mr. Bryan Patterson, Assistant in Paleontology, for assistance and for the table of synonymy.

From the Araucanian formation § (Pliocene) of Paraná, Catamarca, and Monte Hermoso have been described, during the past seventy years, various fragments of lower jaws and isolated teeth of carnivorous mammals. In most cases these specimens have been made types of new genera. Owing to the fragmentary condition of the specimens so described it has been difficult for the present writer to recognize the genera, and in some cases the more general relationships of the forms have remained in doubt. The genera and species of specimens hitherto described from these formations are as follows:

KNOWN GENERA OF BORHYÆNID MARSUPIALS

Eutemnodus americanus Burmeister, 1885.

E. *americanus* Bravard, 1858 (nomen nudum).
E. *americanus* Burmeister, 1885.
An. Mus. Nac. Buenos Aires, Ent. 14, pp. 97–98.
Horizon and locality: Entrerriano (lower Pliocene), Paraná, Argentina.
Type specimen of Bravard redescribed, together with an incisor having similar markings and from the same locality. Referred by Burmeister to the family Felidæ.
Holotype: A molar tooth without roots, described as having peculiar markings in the enamel.

Note: The synonymy of this form has been so confused that a full account of it would be out of place here. A complete table of the synonymy with notes will be found in the appendix to this paper, page 31.

---

* The term "sparassodont" which has been applied by Ameghino to both marsupials and placental mammals is not used in this paper. Borhyænoidea covers the predacious marsupials usually included under it.
† "New Family of South American Pliocene Mammals," *Bulletin of the Geological Society of America*, Vol. 40, p. 117, 1929.
‡ Preliminary Description of a New Marsupial Sabertooth from the Pliocene of Argentina. Field Mus. Nat. Hist., Geol. Ser., Vol. 6, pp. 61–66, 1933.
§ The Araucanian as treated by Rovereto (1914) includes the following stages: Rionegrense, Araucanense, Hermosense and Chapalmalense.

Acrohyænodon pungens Ameghino, 1904.

An. Soc. Cient. Arg., Tomo 58, pp. 267–268.

Figured by Rovereto, An. Mus. Nac., Buenos Aires, 1914, Tomo XXV, p. 150.

Horizon and locality: Hermosense (upper Pliocene), Monte Hermoso, Argentina.

Holotype: Posterior part of ramus of mandible with one tooth and a separate molar of the same ramus.

?A. acutidens Rovereto, 1914.

An. Mus. Nac., Buenos Aires, 1914, pp. 83–85.

Horizon and locality: Araucanense (middle Pliocene), Valley of Santa Maria, Catamarca, Argentina.

Holotype: A fragment of a mandible with last molar tooth.   Doubtfully referred to this genus by Rovereto.

Hyænodonops chapalmalensis Ameghino, 1908.

An. Mus. Nac., Buenos Aires, Ser. 3, Vol. 10, p. 423.

Horizon and locality: Chapalmalense (Upper Pliocene), Province of Buenos Aires, Argentina.

Holotype: A lower molar from which the protoconid and the point of the paraconid are broken away.

Character cited: The presence of a small talonid.

Figured by Rovereto, op. cit., 1914, p. 185.

Parahyænodon argentinus Ameghino, 1904.

An. Soc. Cient. Arg., Tomo 58, pp. 266–267.

Horizon and locality: Chapalmalense (Upper Pliocene), Province of Buenos Aires, Argentina.

Holotype: A hind foot almost complete, accompanied by the canines, an incisor, a first lower molar and other parts of the skeleton—all of one individual (fide Ameghino).

THYLACOSMILUS * Riggs

Genotype: Field Museum No. P14531.

Horizon: Araucanense, Pliocene.

Type locality: Puerta Corral Quemada, Department of Belen, Catamarca, Argentina.

Generic characters: Large carnivorous marsupials having massive face and short cranium; dentition, I_0^0, C_1^1, P_2^2, M_4^4; upper canine teeth strongly developed, deeply rooted and hypsodont; a wide flange developed on the mandible to receive the upper canine; superior branch of the maxilla extending backward between the orbits and meeting at the median line; nasal bones long, attenuate and laterally compressed; orbits entirely enclosed in a bony ring; occipital condyles strongly projecting; a prominent pair of tubercles developed on the basisphenoid bone; entepicondylar foramen and postpalatine vacuities absent; digits V–IV; fore feet digitigrade, hind feet plantigrade.

The skull (Pl. I) as a whole is massive in appearance, presenting a strongly convex facial region, long and trenchant canine teeth, and a short and closely set cranial region.

* Geol. Ser. Field Mus. Nat. Hist., Vol. 6, pp. 61–66, 1933.

The occiput is vertical in profile with condyles protruding beyond the inion. The posterior aspect presents a sub-circular outline with the foramen magnum as a center (Pl. III, Fig. 2). Owing to the great development of the superior branch of the maxilla the orbit appears low on the side of the face. It is unusually small but prominent. The arches are short and broad as in the later machairodonts but less extended beyond the orbits. The sagittal crest is short and high, the inion bears a strong median keel on its posterior surface. The lambdoidal crest is almost vertical to the basal plane of the skull; it is thickened at the periphery, buttressed on its posterior surface and stronger than that of any terrestrial mammal known to the present writer. It divides midway of its length, sending an anterior branch to the base of the zygoma and a posterior branch to the mastoid process. The latter has a development in this animal similar to that of the machairodonts. The auditory meatus opens downward into a narrow cleft. The temporal fossa is sharply defined by the enclosing structures which limit it to a relatively small and rounded outlet below.

The facial region has the peculiarity of overlapping the cranium. It is made up largely of the maxillary bones. The premaxillary and nasal bones are so reduced as to play a minor part in the facial configuration. The frontals appear only in the postorbital processes, in the temporal fossæ, and in the narrow plates which extend backward and inward to meet in the median plane above the anterior margin of the cranial cavity. The lobate superior processes of the maxillary bones contribute chiefly to the convex outline of the face. In places these lobes are so furrowed with canals for the passage of blood vessels, especially in the holotype of *T. lentis* (Pl. IV, Fig. 1), as to suggest that they may have borne a horny covering, though such a development has not been recorded among marsupials. The anterior nares are high and narrow, being crowded between the massive roots of the canine teeth.

The orbit is protected by a strong bony ring to which the frontal, the lachrymal and the jugal bones contribute almost equally. The infraorbital foramen opens anteriorly above the posterior margin of the fourth premolar. The lachrymal duct perforates the lachrymal bone at the anterior margin of the orbit as in the borhyænids. There is an antorbital pit in the maxillary bone which varies in size between the two species hereafter described.

The palate is broad and firmly ossified (Pl. II). It is perforated by a small pair of anterior vacuities and by certain posterior foramina, but there is no trace of postpalatine vacuities. The posterior nares open at a line drawn through the posterior molars as is common in the borhyænid marsupials; the opening is divided by a median septum. The basicranial region presents an unusual feature in a pair of strong tuberosities which arise from the basisphenoid bone midway of its length, and give attachment to muscles which serve to depress the head.

Genotype: T. atrox Riggs.

Species of This Genus

Thylacosmilus atrox * Riggs

Horizon and locality: Araucanense (Pliocene), Corral Quemada, Catamarca, Argentina.

* Geol. Ser. Field Mus. Nat. Hist., Vol. 6, pp. 61–66, 1933.

Holotype: Field Museum No. P14531, skull and parts of skeleton.

Paratype: P14344, cranium, mandible, hind leg and tarsals.

Skull, length in holotype, 260 mm.; P^2 having a single root, P^3 a double root; frontals do not enter into the formation of the sagittal crest.

T. lentis * Riggs

Horizon and locality: The same as *T. atrox.*

Holotype: Field Museum No. P14474, incomplete skull with dentition. Skull, length 197 mm.; P^2 with root grooved indicating inclination to subdivide; P^3 with two roots, the mesial one grooved; frontals enter into formation of sagittal crest.

RELATIONSHIPS TO KNOWN BORHYÆNID GENERA

The genus *Thylacosmilus* includes animals which are much larger than the other Pliocene forms enumerated above (pp. 2–4). Comparison is limited because of the scant material which in most cases composes the types of those genera. It differs from *Eutemnodus* in the absence of "pittings of lace-like pattern" in the enamel of the tooth which have been described as the most distinctive character in the holotype of that genus and which have been recognized in a number of other teeth from the same formation later referred to it (Burmeister and Ameghino). Also, the lower molars of specimens referred to *Thylacosmilus* are distinguished from those referred to *Eutemnodus* by the presence of a talonid. The absence of the talonid as described in the holotype of the latter genus suggests that the tooth in question may be the unworn crown of an upper second or third molar. Nothing can finally be determined in regard to the relationships of *Eutemnodus* until the discovery of other specimens shall make it possible to fix the characteristics and the systematic position of this most uncertain form.

Acrohyænodon was described by its author as a member of the Hyænodontidæ. The genotype, *A. pungens*, is founded upon two isolated teeth which were described as the second and third lower molars of a hyænodont. These teeth are very similar in structure to those of the unworn second and third lower molars of the well-known Santa Cruz genus *Borhyæna*. While the teeth in this species are smaller than those of *B. tuberata*, and there are no means of determining what variations may have existed in other structures of this animal, the two appear to be closely related. No characters so far known should exclude this genus from the typical South American family, Borhyænidæ. The holotype of *?A. acutidens* consists of a fragment of a mandible with a single molar in the alveolus. This tooth has the position of a last lower molar and the structure of a typical borhyænid lower fourth molar. There is sufficient evidence to determine it as such a tooth, rather than the third lower molar of a hyænodont as it was originally described.

A specimen (P14407, Pl. VI, Figs. 4, 5) collected by the Field Museum expedition from the Valley of Santa Maria, Catamarca, the type locality of *?A. acutidens*, has for comparison a broken $M_{\overline{3}}$ and a little worn $M_{\overline{4}}$ in a well-preserved fragment of mandible. The latter tooth is almost identical in measurement and in structure with the holotype of this species. As the two specimens are from the same horizon and locality and are similar in size and in such characteristics as can be determined, it is probable that they are of the same species. The Field Museum specimen is therefore referred to *?A. acutidens.*

* Geol. Ser. Field Mus. Nat. Hist., Vol. 6, pp. 61–66, 1933.

The last molar tooth in both specimens of this species has a vertically directed proto-conid, a considerably lower paraconid, separated from this protoconid by a narrow cleft, bordered by cutting edges. The blade thus formed, especially in the Field Museum speci-men, is set obliquely to the axis of the mandible. A well-developed talonid is present. These are distinctive characters of the more advanced borhyænid marsupials as exemplified in the corresponding teeth of *Borhyæna* and *Prothylacynus* of the Santa Cruz beds. The structure of this tooth is very different from the widely divergent protoconid and paraconid, and the vestigial talonid observed in $M_{\bar{3}}$ of the later members of the Hyænodontidæ. We may therefore fairly conclude that in the two species of *Acrohyænodon* of the South Ameri-can Pliocene formations we have to deal with marsupials of the borhyænid type.

The genus *Hyænodonops* is based upon a single lower molar tooth described by Ame-ghino and figured by Rovereto (1914, p. 185) as a member of the Hyænodontidæ. This specimen is broken at the crown so that little more than half of the protoconid and the paraconid remain. The lower part of the narrow cleft which separates these two cusps appears to be similar in structure to that of $M_{\bar{4}}$ in the borhyænids. The reduced talonid is also similar to the same. However, the base of the protoconid, as figured, shows on the lateral surface a vertical concavity which has not elsewhere been observed in such a tooth. On the whole, the specimen is so poor that it can not be determined with certainty.

Of the genus *Parahyænodon*, Ameghino says in description (1904, pp. 266–67): "These remains are so similar to the corresponding remains of *Hyænodon* that it is difficult to make a generic distinction. The upper canine is almost straight on the posterior margin and curved on the anterior, with the root a little longer than the crown. It has a straight line measurement of 39 mm. [*Translation*]." Only the astragalus of this specimen was figured by Ameghino. Later, all of the bones of the specimen were figured by Rovereto (1914, Pl. XI) and the genus was classified with the Hyænodontidæ by him without further comment. The creodont-like appearance of the astragalus as pointed out by Ameghino appears to distinguish this animal from the known foot of *Borhyæna* and of *Prothylacynus* but possibly not from the foot of *Amphiproviverra* as figured by Sinclair (1906, Pl. LIV, Figs. 6, 8). A careful re-study of the type specimen of *Parahyænodon* in comparison with indigenous and immigrant flesh-eaters will be necessary in order to fix definitely the validity and position of this genus. Upon it appears to rest the sole evidence in support of the claim that members of the Hyænodontidæ reached South America.

The habit of wear in the borhyænines of the Araucanian, as evidenced by three speci-mens of *Acrohyænodon* (the holotypes of two species and a topotype of one of them), is not to abrade the crown. The fact with regard to the one identified tooth of ?*Hyænodonops* can not be determined. But the same habit of wear is apparent among the borhyænines of the Santa Cruz period. Two specimens of *Borhyæna* in the Field Museum collections, one a young adult and the other older, also two specimens figured by Sinclair (op. cit.), show little wear on the lower fourth molar. The same is true of two specimens of *Pro-thylacynus* and one of *Cladosictis* figured by that author, as well as of one specimen of an earlier species of the former genus in Field Museum. It is also significant that in most of these specimens there is evidence of more or less wear on the lateral surface of the lower fourth molar, as a result of shearing action similar to that observed in canids.

The evidence of wear on the teeth of *Thylacosmilus* is quite different. The apices of the crowns and the intervening clefts of the molars are worn down as in a crushing action.

This fact indicates a very different feeding habit in this animal from that observed in the borhyænines. Also, the long and trenchant upper canine tooth of *Thylacosmilus* remains acutely pointed and little worn while the same tooth in the borhyænine forms is blunted and often broken. This evidence leads to the conclusion that the specimens upon which the species of *Acrohyænodon*, and probably also of *Hyænodonops*, are based, are of the narrow-faced borhyænine type of marsupial carnivores and that the genera and species founded upon them should be included in the subfamily Borhyæninæ.

The single tooth used as the holotype of *Eutemnodus* cannot be compared with *Thylacosmilus* in this way. The crown view has not been figured; the crown is apparently unworn. It is quite possible that the type specimen may be an upper second or third molar of an animal related to *Thylacosmilus*.

From the earlier formations of South American Eocene, Oligocene and Miocene ages are known a series of carnivorous marsupials of borhyænid type. They differ from the dasyuroid marsupials of Australasia in the absence of the post-palatine vacuities and in the much reduced protocone of the upper molars. They have been variously credited with close relationships to the Tasmanian thylacines and to the American didelphids.

The species *Arminiheringia auceta* Ameghino (Simpson, American Museum Novitates, No. 578, 1932) is the most completely known borhyænid of pre-Santa Cruz time. It is also one of the largest and strongest of the indigenous South American marsupials. As pointed out by Simpson it is one of the most highly specialized in dental structure. The last pair of molars ($M^{\underline{4}}$, $\frac{}{4}$) were more massive than those of any later known borhyænid. The protocone is much reduced in all of the molars, the metaconid absent, the talonid reduced to a small cingulum-like heel. The shearing edges of the cleft in the last pair of molars are well developed.

Proborhyæna of the Colpodon beds (Oligocene) is known from three fragmentary mandibles and some separate teeth (Ameghino, Secundo Censo de Argentina, 1895. Sinclair, Field Mus. Nat. Hist., Memoirs, Geol. Ser., Vol. I, p. 38). This is a short-faced animal having massive canines, reduced premolars and anterior molars. The fourth lower molar is large and sectorial. This form presents a strong contrast to the long-faced animal of the Notostylops beds mentioned above, as well as to the leading types of the succeeding period. A smaller species of *Borhyæna* and a large one of *Cladosictis* have been recognized from this horizon by Sinclair (op. cit.).

From the Santa Cruz beds (Miocene) are recorded the greatest number of individuals as well as the greatest number of genera and species known to the borhyænid group. Apparently these animals reached their greatest dispersal at that time. As many as twelve genera of them have been described by Ameghino from this horizon. Four genera and seven species were recognized and figured by Sinclair from the collections of Princeton University and the American Museum. The collections of Field Museum from this horizon, studied by Sinclair (op. cit.), have added five specimens to those previously known, but no new species.

Owing to the writer's inability, on the occasion of various visits to Buenos Aires between 1922 and 1927, to gain access to the type specimens of various genera of Santa Cruz marsupials included in the private collections of the late Dr. Ameghino, and owing to the absence in many cases of figures to illustrate the type descriptions, no conclusions are here drawn from Santa Cruz borhyænids excepting those figured in the Princeton University Reports, those by Cabrara (1927) and those included in the collections of Field Museum.

In the Pliocene formations the known borhyænines, or cynocephalous types, as above enumerated, are represented by two or possibly three genera, all of intermediate size. Individuals are so rare as to be known from isolated fragments only. Collectors of the Field Museum expedition, having explicit instructions to be on the alert for any trace of carnivorous mammals, brought in, as the result of six months' search through rich fossil-bearing localities of the Araucanense, only one fragmentary specimen of marsupial belonging to this group. Other fossil-bearing localities of the Pliocene (Araucano) formation, including the Rio Negro, Paraná, and Monte Hermoso, have produced no more than four or five known specimens which may belong to the Borhyæninæ. The paucity of individuals, of species and of genera, would indicate that the cynocephalous borhyænids in this period were much less numerous than in the Santa Cruz. The appearance of placental carnivores of the procyonid, and possibly other families, during the Pliocene period was doubtless making strongly for their extinction.

In the Araucanensian formation of Catamarca were discovered the first known remains of the thylacosmiline group of sabertooth marsupials. These animals were found in such numbers and in such size as to indicate that they were not only the predominating type of carnivorous marsupial, but that they were also holding their own with the large procyonids (*Amphinasua*) which were found there in equal numbers. The habits of these two types of flesh-eaters, so far as may be judged from their structure, were probably so different that they were not brought into close rivalry in their quest for food. The appearance of the canids and the ursids, which occurred at a somewhat later stage so far as is now known, may have played a more important part among the Pliocene faunas. *Thylacosmilus* appeared in the middle Araucanense of Catamarca as an animal entirely unique. Strangely enough there has been recognized among the Oligocene and the Miocene marsupials, which are known from the southern half of the South American continent, no animal which might be regarded as ancestral in a direct way to this form. No other marsupial now known from any part of the world has developed the peculiar weapon known as saber-tooth.

CLASSIFICATION OF THE PREDACIOUS MARSUPIALS

Superfamily	Didelphioidea.
Family	Didelphiidæ, Gray, 1821.
Family	Carloameghiniidæ, Ameghino, 1901.
Superfamily	Borhyænoidea.
Family	Borhyænidæ, Ameghino, 1894.
Subfamily	Borhyæninæ, Cabrera, 1927.
	Borhyæna Ameghino.
	Pharsophorus Ameghino.
	Proborhyæna Ameghino.
	Prothylacynus Ameghino.
	Cladosictis Ameghino.
	Amphiproviverra Ameghino.
	Acrohyænodon Ameghino.
	Hyænodonops Ameghino.
	Etc. etc.
	? *Borhyæninæ* incertæ sedis.
	Eutemnodus Burmeister.
	Parahyænodon Ameghino.
Subfamily	Thylacosmilinæ, Riggs, 1933.
	Thylacosmilus Riggs.
Superfamily	Dasyuroidea.
Family	Dasyuridæ, Waterhouse, 1838.
Family	Notoryctidæ, Ogilby, 1891.

Type and Referred Specimens of Thylacosmilus

The specimens which have here been included under the genus *Thylacosmilus* are three in number. They were collected by the Marshall Field Paleontological Expedition to Argentina and Bolivia in 1926–27. The specimen taken as a holotype of *T. atrox* (P14531, Pl. I) consists of a large skull in which the entire dentition of the right side is preserved, also the canine from the left side, found separate in the matrix a short distance from the skull; fragments of both mandibles found near-by; one entire humerus, the articular ends and part of the shaft of a radius, a pair of femora broken and in fragments, and sufficient separate bones to make it possible to restore a forefoot. Of the skeletal parts, only the humerus remained in the matrix. Other parts had been washed out by a little rivulet which ran down the hillside and had been caught and held in small pockets in its course. From these sources they were gathered up and have been fitted together with great care by the writer and Mr. J. B. Abbott. The missing parts of the skull have likewise been restored; the fragments of the mandible, found scattered and isolated, have been set together and the missing parts restored by using an entire mandible of the smaller paratype (specimen P14344) as a guide. The holotype is apparently a large adult male. It was collected by the writer, assisted by a very efficient peon helper, Juan Mendez. The horizon is the intermediate fossil-bearing level of the Catamarcan formation, designated above as the Araucanense, at Puerta Corral Quemada, Department of Belen, Province of Catamarca.

A second specimen, No. P14344, paratype of the above-named species, consists of a cranium with the roots of the canine teeth in the sockets, a left mandible entire, seven cervical, two dorsal, two lumbar and two sacral vertebræ, a femur, tibia, fibula, astragalus, calcaneum, cuboid and various other foot bones. This is a young adult specimen in which the mandible is nearly one fourth smaller by linear dimensions than that of the holotype. The specimen was found at about the same geological horizon as the first, near San José, Department of Santa Maria, Catamarca, by Mr. Robert Thorne, one of the collectors of the expedition. This is the type locality of the Araucanense formation.

The holotype of *T. lentis* (P14474) consists of a skull in which the entire dentition of the right side is preserved; that of the left side, excepting the root of the canine tooth, has been destroyed by erosion. Judging from the markings of the skull and from the worn dentition, this specimen belonged to an older individual than either of the others. It is about the same size as the paratype of *T. atrox*. It was collected by Dr. Rudolf Stahlecker, a member of the expedition, from the same horizon, and only a few miles distant from the place where the holotype of *T. atrox* was found. By mutual arrangement between Field Museum and the Museum of La Plata, this specimen is to become the property of the latter institution.

The Osteology of Thylacosmilus

The Skull

The premaxillaries are not preserved in the holotype of *T. atrox* but are present in the holotype of *T. lentis*. In that specimen the premaxillaries are short in the antero-posterior dimension. As seen from the palatal surface, they extend forward but little beyond the anterior margins of the canine teeth (Pl. IV). The maxillo-premaxillary suture appears opposite the middle of the canine alveoli. Each bone is cleft posteriorly so as to form a pair of small anteropalatine vacuities. The bones are entirely edentulous in this adult and were doubtless so in the young stage, though no young specimen of this animal is known to the present writer. The superior branches of the premaxillaries form a narrow band about the narial opening and are but little produced at the angle between the nasals and the maxillaries.

The maxillaries are perhaps more strongly developed in this genus than in any other known carnivorous mammal. The outstanding feature of these bones is the great posterior development of the superior processes (Pl. II). These processes extend upward and backward between the orbits, terminating at a point above the anterior margin of the cerebral cavity. They consist of a pair of great, lobelike projections which constitute the alveoli of the upper canine teeth. In their posterior half these processes overlie the nasal bones, concealing them entirely. In so doing, they meet in a simple suture at the median line, from a point above the orbits backward to their posterior extremities. They articulate laterally with the lachrymals and the frontals, and posteriorly with the frontals.

The inferior branches of the maxillary bones have the usual development of those elements in borhyænid marsupials. They are excluded from the rims of the orbits by a small contact between the jugal and the lachrymal bones (Pl. I). On the lateral surface each maxillary is excavated by a wide antorbital fossa which is separated from the infraorbital foramen by a vertical bridge. In the holotype of *T. atrox* this bridge appears to cover a wide foramen which gives outlet to canals leading from the deeper layers of the adjacent bone, but in *T. lentis* no such foramen can be recognized. A smaller foramen in both specimens opens on the facial aspect of the maxillary below the infraorbital foramen and near the alveolar border.

In their palatal aspect the maxillaries are broad, moderately concave on the inferior surface and laterally convex at the alveolar margins (Pl. II). Each bone bears six teeth of the molar-premolar series. The maxillo-palatine suture extends forward to a point opposite the root of the second molar tooth as in *Borhyæna*. A small foramen, common to both the larger and the smaller species of this genus but more conspicuous in the latter, perforates the palatal surface of the maxillary opposite the root of the canine tooth and a few millimeters behind the suture, as in *Borhyæna tuberata*. A second foramen of larger size opens at the maxillo-palatine suture opposite the last molar. While somewhat more anteriorly placed, this is apparently a homologue of the postpalatine foramen in *Borhyæna*. There is no evidence of postpalatine vacuities; in this character *Thylacosmilus* is in agreement with the typical borhyænines.

The nasal bones are long and attenuate and are concealed in the posterior half of their length by the superior maxillary lobes. They are broadest at the anterior end and taper regularly to narrow vertical plates. A natural section in the paratype of *T. atrox* (Fig. 3) shows the nasal bones at a point above the orbits, compressed between the alveoli of the canine teeth. Apparently they retain contact with the frontals below and between the posterior ends of the canine alveoli.

The lachrymal bones (Pl. II) are developed as large triangular laminæ in the anterior walls of the orbits similar to those of *Thylacynus*. The anterior extension of the lachrymal upon the face is limited to a deep fold which forms the anterior margin of the orbit. In addition to this, the lachrymal bone is produced upward and backward between the frontal and the superior process of the maxillary, where it terminates in a narrow point midway between the margin of the orbit and the anterior extremity of the sagittal crest.

The frontal bones (Pls. I, II) are modified from the usual type seen in *Borhyæna* and *Thylacynus* in proportion to the great development of the upper canine teeth and the backward extension of their alveoli. They form the postero-superior third of the orbital enclosure, sending downward a strong postorbital process to unite with the jugal. From this process a stout buttress extends diagonally inward and backward and joins the lateral wall of the cranium. This development effectively separates the superior orbital area from the temporal fossa. The fronto-maxillary suture extends obliquely backward until it comes in contact with the temporal ridge at the extremity of the superior maxillary process. The frontal meets its fellow of the opposite side in a short suture between the termini of the maxillæ and the parietal bones. This median suture, so far as it appears at the surface, has a length of only twenty millimeters in the holotype of *T. atrox* and ten millimeters in that of the smaller species. The lateral plate bounds the anterior half of the temporal fossa.

The jugal bones are short and stout in comparison with those of *Borhyæna* and *Thylacynus*. The anterior end extends a short distance past the anterior margin of the orbit, where it terminates in a narrow point abutting the lachrymal. The maxillo-jugal suture extends irregularly downward and backward below the orbit in a direction similar to that observed in *Borhyæna*, but quite unlike the same in *Thylacynus*. A stout postorbital process of the jugal bone extends upward behind the orbit where it joins the corresponding process of the frontal bone in a firm sutural union. The bar formed by these two processes encloses the orbit posteriorly. The superior margin of this bone forms more than a third of the orbital boundary. The posterior branch forms less than half of the zygomatic arch; it terminates just outside the glenoid cavity in a downwardly curved extremity more like that of *Borhyæna* than that of *Thylacynus* (Pl. I). The superior margin of this branch meets the anterior process of the squamosal in a long S-shaped suture.

The parietal bones (Pl. II) are relatively short antero-posteriorly. They meet at the median line to form a short, stout sagittal crest, to which the frontal bones in *T. lentis* contribute a part. The parietals apparently terminate posteriorly in the massive lambdoidal crest, forming its superior two thirds.

The occipital bones (Pl. III) partake of the massiveness of the entire cranial region. The condyles project well beyond the occiput. The supraoccipital of the holotype of *T. atrox* is marked by strong rugosity at the points of muscular attachment about the superior margin. It bears on its posterior surface a median keel and two strong lateral protuberances. These structures, however, are not so strongly developed in the smaller specimens. The position of the occipito-parietal suture is conjectural. From the structure in the younger paratype specimen of this species, it would appear that the suture closely follows the posterior margin of the lambdoidal crest, but this can not be determined positively. Likewise, the suture between the supraoccipital and the exoccipital bones can not be traced in any of the specimens available for this study. The great strength and rugosity of the muscular attachments developed in the occipital region of these animals apparently has led to closing the occipital sutures early in the life of the individuals. The exoccipital bone overlaps the mastoid process in a suture plainly traceable in the young specimen alluded to above. There is no paroccipital process. The basioccipital is relatively shorter in *Thylacosmilus* than in *Borhyæna*, in *Thylacynus* or in *Dasyurus*. The basioccipital-basisphenoidal suture is opposite the auditory meatus while in the other forms mentioned the suture is opposite the glenoid cavity.

On the latero-inferior angle of the exoccipital, separated from it by a well defined suture, lies the mastoid process.   Whether or not this is the mastoid process of the petrosal, as is observed in a similar position in *Thylacynus*, can not be determined from the specimens in hand.

The basisphenoid bone in *Thylacosmilus* has taken on a more important function than is common among marsupials, in supplying one of the strong muscular attachments for movement of the head.   This attachment consists of a prominent pair of tubercles which arise from the basisphenoid at a point opposite the posterior margins of the glenoid cavities. They may conveniently be designated as the basisphenoid tubercles.   This development has not only expanded the basisphenoid laterally but is probably responsible for its elongation posteriorly.   The function of the tubercles is evidently to give strong attachment to the *longus capitis* muscles as employed in a downward stroke of the head in the use of the great canine teeth.*

The alisphenoid bone (Pl. IV) in its lateral aspect is best observed in the type of *T. lentis;* it is reduced in proportion as the temporal region of the skull has been reduced. A truncated process articulates with the palatal bone opposite the middle of the temporal fossa; a posterior branch extends backward overlapping the auditory bulla.   The lateral process forms the anterior root of the zygoma, extending on its antero-inferior surface as far as the margin of the glenoid fossa.   The superior process extends upward from the anterior root of the arch, and from the posterior angle of the optic vestibule, and terminates in a narrow point between the squamosal and frontal, but has no contact with the parietal. This structure is much modified from that observed in *Cladosictis* and in *Didelphis;* the condition of specimens of *Borhyæna* in Field Museum does not admit of comparisons in this particular.   The orbitosphenoid apparently forms an important part of the mesial wall of the optic vestibule.   On account of numerous small fractures in the younger specimen its outlines can not be traced.

The squamosal bone (Plate I) is preserved entire in the holotype of *T. atrox* only, the cranial portion is preserved in both of the other specimens.   The cranial portion is shortened in its antero-posterior dimension in common with the general shortening of the cranial area.   The glenoid cavity is broad antero-posteriorly, short in the transverse diameter and shallow.   It has only a slight postglenoid process and no trace of the anterior enclosing lip which is observed in *Borhyæna*.   The zygomatic process extends forward in a broad arm which is enclosed by the jugal processes and terminates ten millimeters behind the posterior margin of the orbit.   This S-shaped suture is quite unlike that of either *Borhyæna* or *Thylacynus*.   Both the jugal and the squamosal processes are strong and contribute to the formation of a zygomatic arch, relatively broader and stronger than that of *Smilodon*.

The auditory bullæ (Pl. I) are well developed and prominent.   The composition of the bulla is problematical owing to the closed sutures in the specimens available for study.

---

* As further clarifying the function of the basisphenoid tubercles a study of recent marsupials has been made by Prof. W. K. Gregory in the interest of these researches.   His statement follows: "In response to your letter of September twenty-first, Mr. Raven has just made a sagittal section of a beautifully preserved *Sarcophilus* head.   We find that the *longus capitis* muscles run forward beneath the occipital floor and are inserted in depressions with raised rims.   Comparison with a very large and old *Thylacynus* shows more extended (antero-posteriorly) depressions with high, thick, raised lateral rims provided with a rather prominent tubercle.

"We believe that the extremely high tubercles in your specimen are quite homologous and indicate a further strengthening of the *longus capitis* muscle.   This arises from the sides of the cervical vertebræ.   They are powerful depressers of the head."

A description of the bulla by Mr. Bryan Patterson of Field Museum who has made special studies of the auditory region in South American mammals is appended herewith.

"A meatus spurius is formed by the postglenoid and the post-tympanic processes of the squamosal. The auditory bulla is roughly quadrangular in outline, although differing somewhat in form in the two species; it is flat ventrally and does not extend forward beyond the anterior border of the porus. It is bounded laterally by the very stout post-tympanic process of the squamosal, postero-externally by the mastoid process, which forms a thin plate, closely connected to the post-tympanic process and overlying the bulla to some extent, and postero-internally by the exoccipital which is also plate-like and slightly overlapping. The last mentioned bone does not project freely in the ventral direction.

"There is considerable uncertainty as to the elements composing the bulla. The tympanic seems to have been pinched in between the postglenoid and post-tympanic processes. Deep within the meatus of P14474 this supposed tympanic is visible (Pl. IV) forming the superior wall of the opening into the auditory chamber. Above the tympanic there is a deep dorsal recess of the meatus which leads to a foramen tentatively identified as the subsquamosal. In P14344 the anterior portion of the ventral wall of the bulla is broken away exposing the tympanic which here lies close against the postglenoid process (Fig. 1). The extent of the participation of the alisphenoid in the bulla is not clear. This bone extends posteriorly as a thin strip running between the postglenoid process and the process of the basisphenoid; as far as can be seen in the specimens at hand it extends over the antero-internal corner of the bulla. Whether or not it forms the whole anterior external wall and overlies the tympanic, or whether or not the latter enters into the wall, can not be ascertained. The posterior portion of the ventral wall is probably formed by the tympanic wing of the petrosal but this, again, is uncertain. The participation of the alisphenoid in the bulla of *Thylacosmilus* is an approach to *Cladosictis* and *Amphiproviverra* among the generalized borhyænids rather than to *Borhyæna* and *Prothylacynus* in which no indication of an alisphenoid bulla is present (fide Sinclair)."

Foramina of the Skull

The more important foramina of the skull have been determined from a study of the three available specimens. Comparison has been made with the skull of *Thylacynus* and with specimens of *Borhyæna* and *Cladosictis* of the Santa Cruz. The terminology used by Gregory (Orders of Mammals, 1910, pp. 122 and 221–225) has been followed. A wide variation in the number and in the position of foramina of the skull, as well as throughout the skeleton, has been observed and is treated as a common marsupial character. This variation is particularly apparent among the vascular foramina, many of which are so small and so variable as to be considered as of little value in comparative studies.

The infraorbital foramen is preserved in the holotypes of both species of *Thylacosmilus*. It opens downward and forward at a point midway between the margin of the orbit and the anterior margin of the maxillary, directly above the third premolar. It is much more widely separated from the orbit than the same in *Borhyæna* or in *Cladosictis*. The infraorbital foramen is observed to be widely variable as to position in the different families and genera of marsupial carnivores. In *Thylacynus* it opens at the maxillo-jugal suture while in all other forms observed, both recent and fossil, this foramen is entirely enclosed by the maxillary. With relation to the dental series the position of this foramen is equally variable. In *Thylacynus* it appears above the anterior margin of M³; in *Dasyurus* and in *Borhyæna*, above the middle of M¹; in *Didelphis* and in *Thylacosmilus* it appears above P².

A smaller foramen, particularly conspicuous in the holotype of *T. atrox*, opens at the posterior margin of the canine alveolus below P². Another foramen of secondary importance opens at the antero-inferior margin of the antorbital fossa, leading to a small branching canal. This canal is covered by a bridge of bone eighteen millimeters wide in this specimen. A similar fossa has been described in *T. lentis* but no foramen can be recognized leading from it.

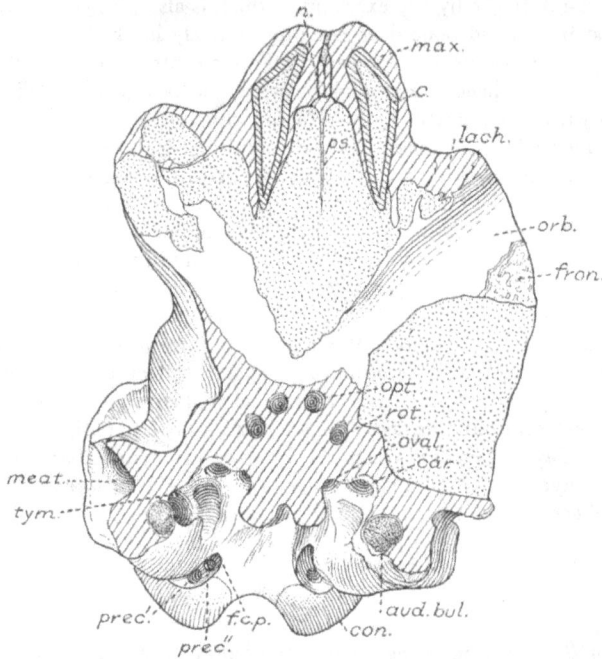

FIG. 1. Natural section of the skull, *Thylacosmilus atrox*. Paratype, P14344. ✕ ¾. opt., foramen opticum; rot., foramen rotundum; oval., foramen ovale. car., foramen carotidum; prec'., foramen precondylum I; prec''., foramen precondylum II; f.c.p., posterior carotid foramen; con., occipital condyle; aud. bul., auditory bulla; n., section of nasal bone; c., section of canine tooth; p.s., presphenoid; orb., roof of orbit; max., maxillary bone; lach., lachrymal bone; fron., frontal bone; tym., tympanic; meat., auditory meatus.

A single lachrymal foramen perforates the orbital plate of the lachrymal bone within the orbit in a position similar to that of *Borhyæna*, but deeper within the orbit than that of *Thylacynus*. Below the lachrymal tubercle is a rounded notch in the margin of the orbit which may or may not give passage to a second branch of the lachrymal duct. The absence of an external lachrymal foramen is observed in this genus as well as in other borhyænids and in dasyurids.

The infraorbital canal perforates the anteromesial wall of the orbit as a vaulted passage twenty-six millimeters high. The spheno-palatine foramen enters the nasal cavity from the mesial angle of this passage at a point opposite the center of the orbit. A considerable variation in the arrangement of foramina of this region is observed between this genus and *Borhyæna*. A specimen of *B. tuberata* in Field Museum shows the spheno-palatine foramen in the position with reference to the orbit as described above, but entirely without the entrance to the infraorbital canal, and fourteen millimeters posterior to it. An earlier species (*B. riggsi*, Sinclair 1930) has the openings of the lachrymal duct, the infraorbital canal and the spheno-palatine foramen, perforating the anterior wall of the orbit in vertical series in the order named.

The optico-sphenoidal foramen, as best observed in the holotype of *T. lentis*, but shown also in Fig. 1 of *T. atrox*, opens at the posterior extremity of a deep fossa, which is enclosed between a fold of the frontal above and the alisphenoid below. At the lateral margin of this fossa, and separated from the last described foramen by a thin partition, is the opening of the foramen rotundum. The arrangement of these two foramina differs from the arrangement in *Thylacynus* in that the fossa that gives outlet to the first is very much deeper and the foramen rotundum is abreast of, rather than behind, the other and larger foramen.

The openings of the basicranial region of these animals are best determined from specimen P14344, Fig. 1. No evidence of a transverse canal has been found in any of the three specimens examined. The basisphenoid region is so modified by the outgrowth of the basisphenoid tubercles that such a canal might well have been eliminated from this genus. According to the observations of Sinclair, as well as those of the present writer, the transverse canal is also absent from the borhyænids.

The foramen ovale and the opening of the entocarotid canal appear side by side at a point opposite the posterior root of the zygoma, and midway between it and the basisphenoid tubercle, but some five millimeters anterior to the basioccipito-basisphenoidal suture. The foramen ovale is the larger of the two and, as seen in the holotype of *T. lentis*, is situated at the bottom of the fossa between the glenoid cavity and the basisphenoid tubercle. The entocarotid canal is directed meso-anteriorly as in *Thylacynus*.

There are two precondylar foramina as in most marsupials, though in some specimens of *Borhyæna* there is but one. The posterior, which is the larger of the two, opens within the margin of the condyle. From this opening a narrow fossa leads forward a distance of sixteen millimeters. At the middle of this fossa the second precondylar foramen opens; the anterior end terminates in the posterior carotid foramen. The jugular foramen has been recognized in the holotype of *T. lentis*, opening beside the posterior carotid. In the more perfectly preserved and less distorted specimen, P14344, this foramen does not appear separately. No stylo-mastoid foramen has been recognized in this genus.

The Mandible

The mandible of *T. atrox* is known from an entire half of that element in the paratype P14344. With this specimen as a guide, the broken fragments from both halves of the mandible of the holotype P14531 have been assembled and the missing parts restored. The former is figured in Pl. IV, Fig. 2, the latter in Pl. I. The description is of course given from the well-preserved paratype.

The mandible in this species is similar in general structure to that of the earlier machairodonts, *Hoplophoneus* and *Eusmilus*. It surpasses them in the direction of its chief specialization, namely, shortening the lever-arm and extending the flange. The coronoid process is shorter, the masseteric fossa is smaller though better defined, the symphysis and the flange are deeper in *Thylacosmilus*. The lever-arm, from the points of muscular attachment to the hinge joint at the condyle, is short and the jaw correspondingly less effective. The whole structure indicates a rather loosely hinged but mobile organ, capable of swinging downward through a wide arc but having comparatively little strength for opposing the upper jaws in the act of seizing. The character of the lower canine tooth, and of the entire molar-premolar dentition, bears out this inference.

The horizontal ramus is straight on the inferior margin, laterally convex at the base of the cheek dentition and laterally concave at the diastema opposite the upper canine. It is broader at the anterior than at the posterior end.

The most extraordinary feature of this mandible is the great downward extension of the symphysis and the development of a wide flange for the reception and protection of the upper canine tooth. This development is proportionately much greater than is found in any of the sabertooth tigers (Matthew, 1910). The lateral surface of the flange is concave, marked by radiating striæ and perforated by no less than eight small foramina. The articular surface at the symphysis shows the same radiating structure. The symphysis extends four fifths of the depth of the flange, leaving a narrow, free margin. The lateral thickness of the mandible at the flange is very slight, giving little of the square chin so characteristic of the machairodonts. The two halves of the mandible were not co-ossified in the younger animal nor do they appear, from the fragments preserved, to have been so joined in the older adult. Two mental foramina perforate the lateral wall of the ramus below the third and fourth premolars, respectively. The dental foramen is rather large, opens beneath the coronoid process and is confluent with the infracondylar fossa. The angle of the mandible is correspondingly shorter than that of the machairodonts and is similar in structure. It is not inflected beyond the mesial surface of the ramus. A deep fossa lies between the condyle and the angle.

DENTITION

The entire upper dentition of the right side is preserved in the holotypes of both *T. atrox* and *T. lentis*. The lower dentition is known from the paratype of the larger species only. The dental formula of *Thylacosmilus* as determined from these specimens is $I\frac{0}{0}$, $C\frac{1}{1}$, $P\frac{2}{2}$, $M\frac{4}{4}$. The outstanding features of the dentition are the total absence of incisors and the extreme development of the upper canine teeth. In the second of these characteristics this marsupial, as above stated, displays a remarkable parallelism with the sabertooth tigers. This similarity has led to the very fitting designation of this pseudo-tiger as the Marsupial Saber-tooth.

The incisor teeth have been eliminated at an earlier stage in the development of this line of animals. A well-preserved pair of premaxillaries in the holotype of *T. lentis* shows no trace of incisors or of incisive alveoli. The same is true of the mandible in the paratype of *T. atrox*.

The upper canine teeth are preserved, in whole or in part, in all of the three specimens described in this paper. In the holotype of *T. atrox* (Pl. I) both canines are preserved almost entire. The right member of the pair is retained in its normal position in the alveolus. The corresponding left tooth had fallen out of the socket but was found near-by and associated with the skull in the matrix (Fig. 2). Fractures through both the right alveolus and the contained tooth made it possible to examine sections of the root while the specimen was being prepared. A natural fracture through the paratype of this species has made a beautiful section (Fig. 3) of both canine teeth in their alveoli at a point above the orbits. In the holotype of the smaller species (Pl. IV) three fourths of the right canine is preserved in the alveolus; the crown of the left has been broken away by erosion, retaining only the root. This wealth of material has given abundant opportunity to determine the entire structure of these teeth.

The upper canine is strongly recurved throughout its entire length. It tapers uniformly from the middle of the inserted portion to the slender and pointed extremity. It grows from a persistent pulp. The posterior end is open and trumpet-shaped, even in the older adult, resembling in this particular the open end of the incisor in rodents. The posterior end extends to within a few millimeters of the extremity of the capsule-like maxillary process which encloses it. The enamel coating extends throughout the entire length of the tooth. The extent of the pulp cavity varies somewhat according to the age of the individual. In the smaller specimen (P14344) the opening extends past the alveolar border and some distance into the exposed portion of the tooth. In the larger specimen a fracture near the alveolar border shows the cavity to be closed. The length of the open end is estimated at ninety millimeters; the entire length of the tooth, measured outside the curve, is 280 millimeters.

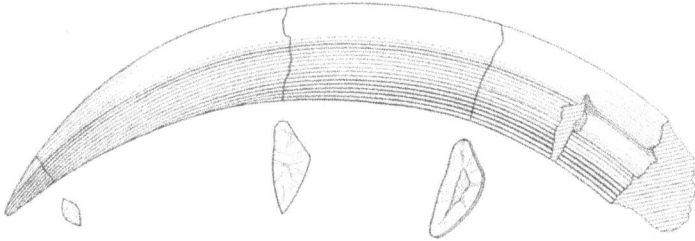

FIG. 2. Lateral view and sections of the left canine tooth of *Thylacosmilus atrox*. Holotype, P14531. × ½.

The enamel coating of the upper canine teeth is very thin. In general it is restricted to the lateral surface, though in the tooth of the larger specimen a thread of enamel extends half way down the exposed mesial surface. At no point is the enamel more than one fifth millimeter in thickness. It is minutely striated in the transverse direction, giving rise to minute denticulations, barely visible to the naked eye where the enamel is exposed at the cutting edge.

The section of the canine (Fig. 2) is sub-triangular throughout the greater part of its length. Near the point it is worn to a lenticular section. The mesial surface grades from a pronounced convexity near the point to a flat surface, rounded at the angles, in the root. The lateral surface is divided into two parts by an angle of 120 degrees. Near the apex of the tooth these two surfaces are equal in breadth; approaching the open end of the upper extremity, the angle inclines obliquely forward so that the posterior surface becomes twice as wide as the anterior one. There is much less evidence of wear on the canines of this animal than upon the corresponding teeth of *Smilodon*. However, the extended pulp cavity and the presence of enamel throughout the entire length of the tooth indicate that the canines continued to grow throughout the life of the individual.   This persistent growth is doubtless responsible for the deep rooting in the superior maxillary, for the backward thrust of that branch in the form of overlying lobes, and for the general modification of the entire facial region of the animal.   The upper canine teeth were the most important element in the entire dental mechanism of this truly bizarre marsupial.

The *molar-premolar* dentition of *Thylacosmilus* is of a marsupial type, closely related to the borhyænids of the Santa Cruz stage.   The premolars are reduced in number and in size to a degree consistent with the overshadowing influence of the sabertooth development in the canines.   The third upper and the fourth lower molars are to some extent specialized as sectorial teeth.   All of the molar teeth have been subjected to wear at the crown so that their original structure can not, in every case, be made out.

The tooth-row of the upper molar-premolar dentition is strongly convex laterally. Between the canine and the next premolar, in the holotype of *T. atrox*, there is a diastema of eight millimeters.   The first premolar is lost; the second is a simple, one-rooted tooth much worn at the crown.   The third is a two-rooted tooth, deeply set in the alveolus and likewise much worn.

The upper molars are all emplanted by three roots, of which the median is much reduced in size.   The crowns of M^1 and M^2 are much worn; the metastyle crest of the former tooth in P14531 is entirely broken away; that of the latter is broad but rounded at the crest.   The protocone of these teeth has been lost; that of *Borhyæna* has been reduced to a vestige.   The paracone is so far reduced as to show only a trace in M^3 of *T. atrox*, the crowns of the preceding molars being so worn as to render the presence or absence of this element indeterminable.   The parastyle present in *Borhyæna* is quite eliminated in both species of this genus.   Molar ^{3} is the strongest tooth of the series and functions chiefly as a sectorial.   The metacone is well developed and much more elevated than in the corresponding tooth of *Borhyæna* or of *Prothylacynus*.   The metastyle crest is strongly developed in M^{1-3}, most strongly in the last.   It joins the metacone in a V-shaped cleft which, in the unworn tooth, is provided with a cutting edge.   The first and second molars in the holotypes of both species are worn so as to destroy most of the cutting edge, but the outline, even in the worn condition, clearly shows the V-shaped cleft.   The last molar is reduced to a strong tubercle with a single stout root and crown consisting of paracone-parastyle joined in a short transverse crest.   It is inserted by a single root, but a groove on the postero-internal surface indicates the recent fusion of two.   The anterior margin of the crown is abraded by the shear of the opposing lower molar.

In general, it may be said that the upper premolar dentition of *Thylacosmilus* is much more reduced than that of *Borhyæna* of the Santa Cruz stage, while the upper molars have a more elevated metacone and are more advanced, especially M^3, toward the sectorial type.

The *lower dentition* of *Thylacosmilus* is known from P14344 (Fig. 3).   This specimen is a young adult as is indicated by the worn canine and molars in connection with the open sutures observed in the cranium.   There is no trace of an alveolus for lower incisors, as the symphysis is crowded closely against the root of the canine.   Those teeth were apparently eliminated at a much earlier stage of specialization in this group of animals.

The lower canine is a strong tooth, oval in cross section and extending 15 millimeters above the alveolar border.   It has an antero-posterior diameter of 11 millimeters and a transverse diameter of 6 millimeters at the root.   The crown in this specimen is worn and blunted leaving no trace of enamel. A diastema of 25 millimeters separates the canine from the next tooth of the series. The *first premolar* has been eliminated entirely; the second and third premolars are simple conical teeth, small and each inserted by a single root. Their crowns are covered with enamel to the extent of 3 millimeters and are but little worn.

The first and second lower molars are worn away at the crown so that their original pattern can only be conjectured. A shallow ring of enamel remains about the base. The crowns of these teeth are much elevated above the alveolus, leaving the roots exposed. Each of them is inserted by two roots of which the anterior one is a little larger. The third molar is a trilobate tooth with moderately elongated crown. The protoconid and the paraconid are worn and rounded in this specimen but apparently were strongly diver-

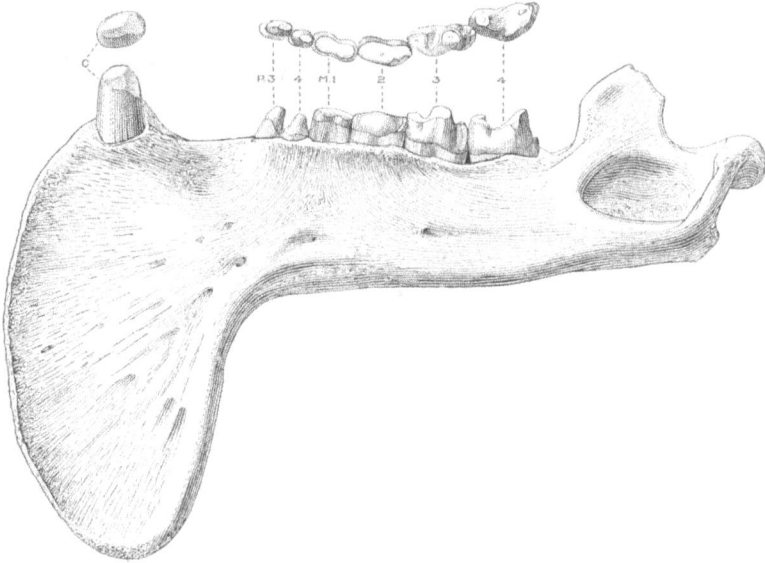

Fig. 3.   Lateral view of mandible of *Thylacosmilus atrox*.   Paratype, P14344.   × ¾.

gent in the unworn tooth. The talonid is more strongly developed than that of the corresponding tooth of *Borhyæna tuberata*. The anterior root is much stronger than the posterior one. Evidence of a deep cleft between the protoconid and the talonid indicates a tendency toward developing a shearing blade.

The fourth lower molar is a sectorial tooth worn and abraded at the crown. The cleft between the protoconid and the talonid is worn so that in the present condition its margins form a right angle. The talonid is reduced to a vestige as is true of the same element in the corresponding tooth of *Borhyæna*. The most evident difference between the two appears in the outline of the crown at its base and in its position in the tooth-row. The outline in *Thylacosmilus* is less convex, and the position is parallel with the ramus.

The borhyænids, while in most forms retaining the primitive premolar formula, show a general trend toward developing a sectorial type of dentition. The sectorials are best represented by the unworn upper tooth of *Thylacosmilus* and by the lower fourth molar of *?Acrohyænodon acutidens* as shown in the referred specimen P14407 (Pl. VI, Fig. 5). The earlier stages of this development are well known. *Amphiproviverra* has a primitive upper molar similar to that of the didelphids, in which the protocone is functional, the paracone and metacone are prominent and of almost equal size, and the parastyle is well developed. In *Cladosictis* and *Prothylacynus* the protocone, paracone * and parastyle are reduced and the metastyle crest is increased in size. In *Borhyæna* the protocone and the paracone are greatly reduced. *Thylacosmilus* carries these steps still farther in the loss of the protocone and the parastyle, and in the extreme reduction of the paracone and the increase in length of the metacone and of the metastyle crest.

Mechanical Adaptations in Skull and Mandible

The extreme specialization observed in the skull and mandible of *Thylacosmilus* has evidently resulted from a series of adaptations in animals of this line. The moving factor

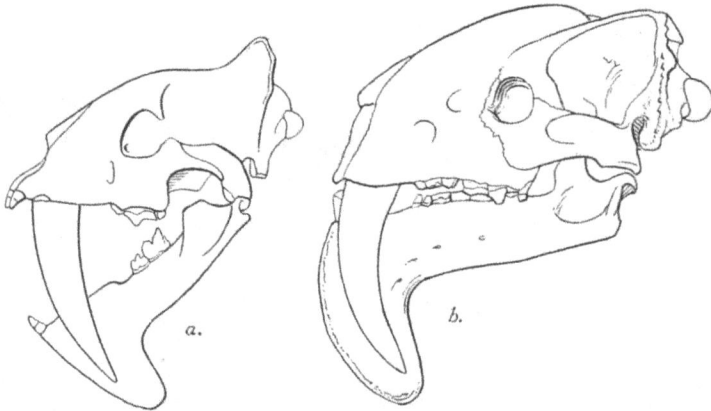

Fig. 4. Comparison of the skulls of *Eusmilus sicarius* (a) and *Thylacosmilus atrox* (b), drawn to the same scale.

in these adaptations has been the extraordinary development of the upper canine teeth, accompanied, no doubt, by highly specialized preying habits acquired by the animal. Great stress was being laid upon the canine. As a corollary of the development of these weapons, there are observed evidences of great relative strength in the vertebræ of the neck and in those muscular attachments of the head which give insertion to muscles controlling its movements. This is especially apparent in the strength of the attachments at the occiput, at the basicranium, and at the ventral surface of the centra of the cervical vertebræ. While all these attachments give evidence of great strength, the series at the ventral surface of the cranium and at the ventral processes of the cervical vertebræ, which

* On M^4 of all borhyænids the paracone and parastyle are retained, forming a transverse crest. Traces of the metacone occur in M^4 of specimens of *Amphiproviverra manzaniana, Cladosictis* and *Prothylacynus* (Sinclair, 1909).

give insertion to muscles used in forcing the head downward, have an unusual development. This fact strongly indicates that they must have been used in a downward stroke of the saberteeth, or in retaining a hold secured by those weapons.

As the crowns of the upper canine teeth, in process of development, became more and more elongated, a deeper rooting naturally resulted in order to give the necessary strength. Pushed on by the backwardly growing root, the superior branch of the maxillary bone was more and more extended in order to accommodate the socket of the tooth. This is evident from the varying development of the canine between the two species. As a result of this extension, one third of the total length of the tooth in *T. atrox* lies back of the center of the orbit. A similar condition has resulted in the machairodonts, but in that group the specialization has not been carried so far. The canine tooth in such Oligocene forms as *Hoplophoneus primævus* had not pushed the extension of the root beyond the anterior third of the orbit. In the later and more highly specialized forms, *Machairodus* and *Smilodon*, the root of the canine has extended backward to a point above the center of the orbit; the suture had apparently become fixed at that point. The root of the canine (cf. *Smilodon californicus*) in adult machairodonts is closed and its development fixed, while in *Thylacosmilus* the extremity of the root is open and the tooth subject to growth throughout the life of the individual. The backward extension of the root might therefore have continued to an even greater length than is observed in these specimens.

Various other modifications may be considered as incidental to the tooth development but essential in supporting the acquired function. The deep rooting of the canine and the backward extension of the maxillary have resulted in a great backward extension of the lachrymal bone, so that more than half of its length lies above and behind the center of the orbit. The nasal bones have been elongated and crowded together at the median line by the overthrust of the maxillary lobes. The frontals have been thrust apart and deformed by the maxillary lobes entering as a wedge between them. The cranium has been shortened in an evident effort to apply the muscular force more nearly to the hilt of the canine weapon. The incisor teeth, both above and below, have been quite overshadowed and eliminated by the great canines. Closing the orbit served to give additional strength to the cranium by furnishing strong lateral buttresses between the cranium and the shortened and greatly strengthened zygomatic arches. It also gave additional protection to the eye. The enlarged and thickened occipital crests, the mastoid and basisphenoid attachments completed a strong circle of muscular insertions about the occipital condyles. The protruding position of the condyles gave added length to the neck muscles and increased freedom of movement to the head.

The mandible of *Thylacosmilus*, again assuming that this form is derived from a typical cynocephalous borhyænid, has undergone modifications no less extreme than those to which the skull has been subjected. A striking parallelism with the structure observed in certain machairodont felids is again found here. This parallelism consists in: a short lever arm from hinge to point of muscular attachment, a low and vertically directed coronoid process, a short but deep masseteric fossa, an infracondylar fossa, a deep symphysis, and a great descending flange which served to receive and protect the trenchant canine tooth.

The construction and free attachment of the mandible in the machairodonts, so as to admit of swinging downward through a wide arc, has been pointed out by Matthew and others as a structure which materially contributed to the efficient use of the great canine tooth in striking with the mouth open. In *Thylacosmilus* the evident capacity for such use of the canine is even more pronounced. The low-hung point of attachment, the simple articulation of the mandible, the operation by muscles attached to a short lever-arm, all indicate a great freedom of action.

Accompanying the extreme development of the upper canine tooth there is also a much greater development of protecting flange than is known in any of the machairodont saber-tooths. The entire suppression of the lower incisor teeth and the elimination of the incisive alveolus is another result of the stress laid upon the upper canine. The reduction of the premolar teeth has apparently followed from the same cause, and indeed the degeneracy in use of the molar dentition, as evidenced by abrasion at the crown, appears as evidence of a habit which no longer relied upon shearing teeth, although the cheek-teeth are of a shearing pattern.

Whether or not the peculiar specialization in the mechanical construction of *Thylacosmilus* would have gone on to even greater extreme had not the appearance of placental mammals in South America introduced as a competitor a new and more highly organized type of carnivorous mammal, can only be conjectured. This new competition evidently resulted in the elimination of, not only the marsupial sabertooth, but the entire family of borhyænid marsupials.

VERTEBRÆ

The entire series of cervical vertebræ, the first and fifth dorsal, two posterior lumbar and two sacral vertebræ, are preserved in the paratype of *T. atrox*. Parts of the atlas and axis of the holotype of this species are also preserved. The cervical vertebræ of the paratype with the first dorsal were articulated in the matrix. The entire cervical series is marked by strong muscular attachments. This is true of the spinous and transverse proc-

MEASUREMENTS

	Specimen No. P14531	Specimen No. P14474	Specimen No. P14344
Skull			
Length, premaxillaries to condyles.............	260 mm.	197 mm.	
Length, margin of canine to post. nares.........	128 "	102 "	
Length of basicranium......................	72 "	55 "	
Condyle to crest sph. tubercle...............	60 "	48 "	
Breadth of arches.........................	176 "	112 " 1	
" " condyles.......................	60 "	56 "	56 mm.
Length of dental series.....................	125 "	101 "	
Length of canine tooth outside curve, as figured .	285 "		
Length of canine tooth beyond alveolus........	124 "	109 " 1	
Greatest breadth of canine..................	35 "	28 "	
" thickness of canine..................	14 "	11 "	
Length of diastema........................	7 "	6 "	
Length of molar-premolar series..............	79 "	67 "	
Molar series..............................	50 "	61 "	
Breadth of M³ at crown.....................	19 "	16 "	
Greatest length of maxilla..................	196 "	139 "	
Length of sagittal crest.....................	44 "	38 "	
Antero-post. diameter of orbit................	27 "	32 "	
Vertical diameter of orbit...................	31 "	35 " 1	
Greatest breadth across mastoid processes.......	99 "	80 " 1	84 mm.

¹ Estimated.

Measurements (*Continued*)

	Specimen No. P14531	Specimen No. P14474	Specimen No. P14344
Mandible			
Axial length, projected.....................	216 mm.[1]	·	175 mm.
Depth at flange.	127 " 1		109 "
Length, condyle to ant. margin canine alveolus..	203 "		152 "
Depth of symphysis........................	104 " 1		86 "
Antero-post. diameter........'.............	57 " 1		44 "
Length of diastema........................	36 " 1		25 "
Length of molar-premolar series..............	80 "		64 "
Length of lower canine.....................			27 "
Breadth of lower canine at alveolus...........			11 "
Breadth of lower fourth molar................			18 "
Depth of ramus at first molar................			34 "
Depth over coronoid process.................	58 "		47 "
Distance, last molar to condyle..............			53 "
Length of masseteric fossa...................	55 "		34 "
Transverse breadth of condyle...............			36 "

[1] Estimated.

esses as well as of certain ventral processes, both single and paired, which have unusual development in this animal. The dorsal vertebræ, so far as known, are similar to those of *Borhyæna;* the lumbars have low arches and broad, low spines.

The atlas has a structure similar to that of the machairodont saber-tooths (Pl. VIII, Figs. 1, 2). It is characterized by the long antero-posterior diameter of the arch, and by the oblique, backward direction of the transverse processes. There is no intercentrum. The anterior articular facets extend to a level with the neural arch, but are separated by a distance of twelve centimeters at their inferior extremities. The posterior facets approach each other more closely at the median line. There are no vertebrarterial foramina. The atlas is pierced laterally by two foramina. The anterior and larger of the two has the usual position of the atlantar foramen in borhyænoid marsupials, above the anterior margin of the transverse process. The smaller foramen is present in one side only and appears above the base of the process.

The axis is more nearly of the borhyænid type, as represented by *Borhyæna* and *Pro-thylacynus,* than is the atlas. The centrum is one fourth longer than it is broad; the spine has the halberd-shape described in *Borhyæna,* is low at the anterior extremity and overhangs the short odontoid process. The posterior extremity of the spine is broken and lost from this specimen. Its thinness at the point of fracture would indicate that the posterior process was not so strong or so much produced as that figured by Sinclair in *Borhyæna.* A strong ventral process, similar to that figured in the above mentioned form, arises from the posterior half of the centrum. The axis is perforated by a larger number of foramina than is found in the better known borhyænids. In addition to the vertebrarterial foramen which perforates the base of the transverse process, a second foramen of similar size opens above the base of that process. Also, a short canal enters the base of the pedicle, at the anterior margin just above the articular facet, traverses the pedicle a distance of six millimeters on the right side of the bone and twice that distance on the left side, and reappears above the anterior margin of the transverse process.

The remaining cervical vertebræ (Pls. V, VI) are somewhat larger and stronger than those of *Borhyæna tuberata* as indicated above, and vary from them in details only. The centra are strongly keeled and ventral processes are developed on the third, fourth and fifth. In the latter the ventral process is bifurcate posteriorly. The vertebrarterial canal perforates the bases of the transverse process of cervicals II to VII as is common among marsupials. In cervicals III and IV the anterior branch of the transverse process is closely applied to the centrum, the posterior or pleuraphysial plate is much more elongate than the same in *Borhyæna*. In V the anterior branch is extended laterally and forms a prominent muscular attachment; the posterior branch is directed outward and upward. In VI the inferior lamella attains the usual development, but is elongated posteriorly and terminates in a strong process for muscular attachment. The superior process, or diapophysis, is thickened and tubercular. Cervical VII has a simple transverse process, curved outward and slightly upward and perforated at the base by the vertebrarterial canal. While the superior branch of the transverse process is developed in *Borhyæna* and perhaps in other of the borhyænids (Figs. 2, 3, Pl. LII, Sinclair) this process in *Thylacosmilus* is much more strongly developed in cervicals IV, V and VI than has been observed in any of these forms.

The neural spines throughout the series are relatively short and broad. That of the third cervical is directed backward, while the fourth is vertical and tapering, the fifth and sixth are inclined increasingly forward, and the seventh is broken in this specimen but apparently was inclined forward. The spines, therefore, form an anticline with the crest at the middle of the cervical series as in *Borhyæna*. The epiphyses of the centra are separate throughout this series and are annular as in the latter genus.

Of the dorsal vertebræ (Pl. V, Fig. 1) but two are preserved and those belong to the same specimen as the cervical series. The first has a centrum with a slight keel as in *Borhyæna* and a tapering spine of moderate length, a low but strong neural arch, short, stout transverse processes, capitular and tubercular facets facing downward. Another dorsal vertebra, estimated as the fifth, has a tapering spine and a short transverse process with rib facet facing downward and outward.

Of the lumbar vertebræ (Pl. V, Fig. 4), the posterior two are preserved in articulation and joined to the sacrum in series. The centra are more rounded than those of *Thylacynus*, the neural arches are low and broad, the spines are broad at the base and relatively low. The facets of the zygapophyses present meso-laterally and interlock; a short metapophysis rises above the articulating facet. A pair of anapophyses are present on the anterior vertebra of the two. The arches are perforated by three or four foramina each, the centra by one or two small foramina on the inferior surface.

The sacrum, as is common among the Borhyænidæ, is composed of two vertebræ. They are firmly co-ossified by the centra and by the arches. The two neural spines are distinct and well developed. The transverse processes are joined by their distal ends; those of the anterior vertebra are much stronger and form the chief contact with the ilium. The distal end of the process, in the second vertebra, remains distinct. The size of the posterior central facet and the well-developed postzygapophyses and sacral spines indicate a moderately long series of caudal vertebræ.

Only one caudal vertebra ia preserved and that in the holotype of *T. atrox*. It is moderately elongated and may have been third or fourth in the series. It has an open neural canal and broken surfaces which indicate the presence of strong transverse processes.

MEASUREMENTS

	Specimen No. P14531	Specimen No. P14344
Vertebræ		
Atlas, breadth across anterior articulation....................	66 mm.	55 mm.
" " " transverse processes...........................		106 "
" " of inferior arch.............................	23 "	19 "
" " " superior arch.............................	43 "	37 "
Axis, length of centrum with odontoid process.................	77 " 1	63 "
" breadth across anterior articular end....................	51 "	45 "
" length of spine.......................................		92 " 1
" breadth base of pedicles..		40 "
Cervical 3, length of centrum...		36 "
" " height, ventral process to crest of spine.....................		72 "
" " breadth of postzygapophyses..............................		40 "
Cervicals 5, 6, 7, length of three centra...............................		98 "
Cervical 6, breadth across transverse processes........................		62 "
" 6, height, base of centrum to crest of spine.....................		64 "
" 7, breadth of transverse processes...........................		83 "
Dorsal 1, length of centrum..		28 "
" " height, base of centrum to crest of spine.....................		74 "
" " breadth across transverse processes.........................		80 "
Lumbar 6, length of centrum..		34 "
" " height, base of centrum to crest of spine.....................		67 "
Sacrum, length of two co-ossified centra.............................		53 "
" breadth across processes at anterior end........................		59 "
" " " anterior end of centrum........................		31 "
" " " posterior end of centrum........................		21 "
First rib, length, tubercular facet to distal end.........................		70 "

1 Estimated

THE LEGS AND FEET

The humerus (Pl. VI, Fig. 1), as preserved in the holotype of *T. atrox*, is a relatively short and stout bone with a massive deltoid area, a strong supinator ridge and small internal condyle. There is no trace of an entepicondylar foramen. The articular surface of the head is oblique to the shaft. The extremity of the great tuberosity has been destroyed in this specimen. The lesser tuberosity terminates some distance below the articular surface. The bicipital groove is broad and shallow as in the machairodonts. On the anterior margin, at the lower extremity of the deltoid area, and well below the middle of the shaft, is a prominent tuberosity for muscular attachment not observed in any other of the borhyænids. The supinator ridge terminates in a tuberosity similar to that observed in *Prothylacynus*. The epitrochlear fossæ do not perforate the bone. The humerus shares the general characteristics of the crouching animals, as distinguished from the cursorial, in the sturdiness of the bone as a whole and in the great strength of the deltoid crest and the supinator ridge.

Of the radius (Pl. VIII, Figs. 3, 4), the extremities and the lower fourth of the shaft are preserved in the holotype of *T. atrox*. This bone appears to have been almost as large as the tibia. The articulation for the humerus is oval in outline and deeply concave; that for the ulna extends more than half way round the head, indicating great freedom of rotation. The articulation for the scaphoid is irregular in outline and does not cover the entire distal end of the bone. It is deeply concave from side to side but only slightly so in the antero-posterior direction. A small facet for the distal end of the ulna faces proximo-laterally. The lower portion of the shaft is convex on the anterior surface, flattened on the posterior one.

The femur (Pl. VII) is represented by both right and left bones of the holotype, *T. atrox*, and by the left one in the younger paratype. The former specimens have suffered from erosion at the proximal ends but are continuous in the shaft; the left member of the pair has most of the distal end preserved. The femur of the paratype has the extremities well preserved but has lost some 20 millimeters of the shaft. The missing portion has been restored by comparison with the paratype.

The femur is relatively straight in the shaft as compared with other marsupials; the great trochanter is low and the head moderately constricted. The head is well rounded as in the machairodonts rather than as in the borhyænines. The articular surface extends well downward forming two thirds of a sphere, with the pit for the ligamentum teres near the center. The digital fossa lies farther to the lateral margin of the shaft than in any borhyænid so far observed. From the lateral view this pit is only partly concealed by the great trochanter. The upper extremity of the fossa terminates on the lateral surface just below the crest of the great trochanter. The great trochanter in the older specimen of *T. atrox* extends downward in a rugose crest, convex in profile and in marked contrast to the concave profile of this part of the femur in *Borhyæna*. The second trochanter is only a little less prominent than the first. The linea aspera are not well marked. The outer and inner condyles are similar in size and separated by a relatively narrow intercondylar notch. The articular surface for the patella is broad and shallow as in *Borhyæna* and like it extends obliquely toward the lateral margin of the bone.

A single tibia (Pl. VII, Figs. 1–3) from the paratype of *T. atrox* is approximately three fourths as long as the femur of the same individual. It is relatively large at the proximal end, tapering in the upper two thirds of the shaft and curved backward in the lower third. The external malleolus is broad and rounded as in the borhyænines. The articular surface for the astragalus is strongly concave from side to side but nearly plane antero-posteriorly. The lateral surface presents a rounded articulation for the fibula. The internal malleolus is stout and bears half of the articulating surface for the astragalus.

The fibula is known from a single bone associated with the tibia last described. The shaft is irregularly rounded in form, being broader laterally throughout the proximal two thirds, but broader antero-posteriorly in the distal third. The distal articulation is sub-cordate in outline with the larger lobe extending well over the mesial surface as in *Pro-thylacynus* (Sinclair, 1910, Pl. LI). The articular surfaces for astragalus and calcaneum are approximately equal, that for the former is convex, for the latter concave.

The forefoot of *Thylacosmilus* is pentadactyl and apparently digitigrade. It is similar in a general way to that of *Borhyæna tuberata*. Evidence of the digitigrade position is derived from a plano-convex articulation between the radius and the scaphoid. The latter bone bears an articular surface, the greater part of which is nearly plane in the antero-posterior direction, although the antero-superior angle is rounded so as to admit of a limited forward movement. The cuneiform bone presents to the ulna a concave, pit-like surface and is capable of a forward and backward movement only.

The left forefoot of the holotype of *T. atrox* has been reconstructed from elements of both right and left feet found disarticulated and washed out on the surface of the ground. In this reconstruction all of the important bones except the trapezium and the trapezoid and the fourth metacarpal have been available. The scaphoid and the pisiform are known from the right foot; the lunar, the cuneiform and the magnum are present in both sides, the unciform from the left side only. The first metacarpal, the distal half of the second, both extremities of the third and the distal end of the fifth, together with seven phalanges, were recovered. From these parts, and by means of careful comparison with the manus of *Thylacynus* and with Sinclair's figures of *Borhyæna*, the foot has been reconstructed with great care by J. B. Abbott, veteran preparator in paleontology. The length of the metacarpals being in doubt, the proportions of the last named genus have been followed as a guide.

The scaphoid is the largest element of the carpus and is proportionately larger than that of *Borhyæna*. The radial surface is strongly convex laterally; it extends over one third of the anterior surface. This bone presents a lateral facet to the lunar, a distal one to the trapezium and trapezoid, and a latero-distal surface to the magnum. The lunar is a relatively small, narrow bone, presenting a rounded surface to the radius, a small crescentic facet to the cuneiform and a saddle-shaped surface to articulate equally with the magnum and the unciform. The cuneiform articulates with the ulna by a rounded and concave facet, and with the pisiform by a somewhat smaller surface; distally it presents a concavo-convex facet to the unciform, narrowly escaping contact with the fifth metacarpal. The pisiform is a short, stout bone, having an irregular concavo-convex articulating surface of which the lower half meets the unciform, the upper half the ulna.

The distal row of carpal bones articulates with metacarpals II to V in an almost straight line. The os magnum differs from that of *Thylacynus* in articulating distally with metacarpal III only. It is closely articulated with the unciform. The latter is the largest bone of the distal row and is supported by metacarpals IV and V.

The first metacarpal is short and stout, with distal articulation slightly oblique to the long axis of the shaft. The second is figured from the distal half only. The third, judging from the known articular ends, appears to have been almost as strong as the second. The fourth is entirely reconstructed from comparisons with *Borhyæna* and *Thylacynus*. The fifth has the distal articulation preserved but is otherwise reconstructed from the opposite bone of the same specimen. Seven phalanges used in reconstructing the foot have been selected from the disarticulated pieces belonging to this specimen. Their positions have been determined from comparative study only. The proximal half of a single ungual suggests that the structure of the unguals was probably similar to those of *Borhyæna*. The foot as a whole is one fifth longer than that of *B. tuberata*.

The structure of the hind foot is less completely known in this genus than is that of the forefoot. The following description is based upon elements preserved in the paratype of *T. atrox*. Weathered fragments of the astragalus and calcaneum of the holotype were also available. In the former specimen the astragalus and calcaneum were found in association in the matrix with the tibia, fibula, cuboid, ectocuneiform and the vestigial mt. I. There can be no question that all of these parts belong to one and the same individual. The structure of all of these elements is so similar to those of *Prothylacynus patagonicus* as to indicate a close relationship; the hind foot of *Borhyæna* has not been figured and so is not available for comparison.

The astragalus in *Thylacosmilus* is but little longer than it is broad. The head is directed forward and is not separated from the body by a neck. The articulation for the tibia is almost equally divided between the proximal and the mesial surfaces; that for the fibula covers the lateral one third of the proximal surface. The articulation for the tibia extends within six millimeters of the anterior end. The facet for the navicular is strongly convex in the vertical direction but approximately plane in the transverse. The facet for the sustentaculum is likewise strongly convex; it covers the latero-inferior surface of the head and extends backward to the center of the inferior surface. A deep fossa separates the latter from the elongate and concave ectal facet. In the plantigrade position of the foot, the astragalus has its long axis directed downward at an angle of twenty-five degrees to the horizontal.

The calcaneum is relatively short, concave on the distal end and thickened at the tuber calcis. Its length is divided into two almost equal parts by the projecting posterior margin of the astragalar facet. It has a rounded ectal facet, a strongly projecting sustentaculum and a concave and moderately oblique facet for the cuboid.

The fibula articulates by its truncated distal end with the lateral margin of the ectal facet of the calcaneum as well as with the proximo-lateral surface of the astragalus.

The cuboid is only a little longer than it is broad. It presents an oblique and strongly convex surface to the calcaneum, a plane facet suboval in outline to the navicular and the ectocuneiform, and a moderately concave facet to the fourth and fifth metatarsals. In general it is quite similar in proportions to that of *Borhyæna*. A rudimentary first metatarsal preserved with this specimen has an oblique facet at the proximal end and tapers to a vertically flattened and rugose extremity. This bone is very similar to that figured by Sinclair in *Cladosictis*.

MEASUREMENTS (*Continued*)

	Specimen No. P14531	Specimen No. P14344
Legs and feet		
Humerus, greatest length............................209 mm.		
" breadth across head and tuberosity.................. 64 "		
" " " distal articulation................... 45 "		
" " " condyles.......................... 66 "		
Radius, diameter of proximal articulation.................19 × 27 "		
" " " distal articulation...................21 × 36 "		
Femur, length in axial direction............................253 "	220 mm.	
" breadth over head and trochanter.................... 58 "	54 "	
" greatest breadth of distal end...................... 54 "	48 "	
" diameter of shaft at middle.......................... 25 "	21 "	
Tibia, length, spine to malleolus....................................157 "		
" breadth, proximal end...................................... 45 "		
" antero-posterior diameter..................................... 46 "		
" breadth, distal end........................: 24 "		
Fibula, length....................................148 "		
" greatest diameter at distal end................................ 20 "		
Fore foot, length, scaphoid to tip of ungual, third digit.........127 " [1]		
" " scaphoid, greatest breadth....................... 26 " [1]		
" " pisiform, length.......................: 24 "		
" " breadth, distal row of carpals...................... 49 "		
" " length, first metacarpal.......................... 26 "		
" " " third................................. 51 " [1]		
" " " fifth................................ 42 "		
" " " first phalange, digit III.................... 18 "		
" " " second phalange, digit III.................. 13 "		
Hind foot, astragalus, length.. 28 "		
" " " breadth of superior face.......................... 22 "		
" " calcaneum, greatest length.............................. 47 "		
" " " " breadth............................... 26 "		
" " cuboid, length, 18; breadth, distal end...................... 17 "		
" " vestigeal metacarpal, length.............................. 27 "		

[1] Estimated.

COMPARISON OF THYLACOSMILUS WITH SANTA CRUZ BORHYÆNINES OF THE GENERA BORHYÆNA
PROTHYLACYNUS AND CLADOSICTIS

Thylacosmilus	*Santa Cruz Borhyænines*
1. Dentition I_8^0, C_1^1, P_2^2, M_4^4.	1. I_3^{4-3}, C_1^1, P_3^3, M_4^4.
2. Canine teeth hypsodont.	2. Canines have closed pulp cavity in the adult stage.
3. Palate broad posteriorly; no post-palatine vacuities.	3. Palate of intermediate breadth; no post-palatine vacuities.
4. Line of molar-premolar dentition strongly convex laterally.	4. Line of molar-premolar dentition laterally concave at anterior end, convex posteriorly.
5. Length of basicranial area (margin of condyles to glenoid articulation) equals one-fourth total length of skull.	5. Length of basicranial area equals one-fifth of total length of skull.
6. Orbit firmly enclosed posteriorly.	6. Orbit open posteriorly.
7. Nasals long, narrow and concealed posteriorly.	7. Nasals are of typical marsupial proportions, expanded distally.
8. Maxillaries are excluded from contact with frontals.	8. Maxillaries excluded from contact with frontals by expanded nasals.
9. Basisphenoid is perforated by an entocarotid canal.	9. Basisphenoid is perforated by entocarotid canal common to marsupials and monotremes.
10. Auditory bullæ present and moderately inflated.	10. Auditory bullæ present in some genera, absent in others.
11. Jugal bar terminates just outside the glenoid cavity.	11. Jugal bar enters into formation of the glenoid cavity.
12. Zygomatic process of the squamosal terminates directly below the post-orbital process of the jugal.	12. Zygomatic process of the squamosal terminates near the middle of the arch.
13. Lachrymal bone extended to form the anterior margin of the orbit and produced upward and backward beyond the orbit. Perforated within the orbit.	13. Lachrymal bone extended on the face in front of the orbit, perforated within.
14. Mandibular symphysis vertical, deep and develops a flange.	14. Symphysis shallow, procumbent and without flange.
15. Mandibular angle slightly inflected but not extended mesially beyond the dental foramen.	15. Mandibular angle strongly inflected as in Dasyuroidea.
16. Basisphenoid bears a pair of prominent tubercles for the insertion of strong longus capitis muscles.	16. Basisphenoid muscles bear a rugosity at this point. Tubercles present in some old individuals of *Thylacynus*.
17. Postzygomatic foramen not present, postglenoid foramen present.	17. Postzygomatic foramen, in addition to postglenoid, perforates posterior surfaces of the zygoma.
18. A vascular foramen perforates the squamosal within the external auditory meatus, external foramen not in evidence.	18. A large vascular foramen perforates the squamosal above the opening of auditory meatus.
19. Sutures of the facial region of the skull and the zygomata, of vertebræ and of long bones, are open in adult specimens. Those of cranial and occipital regions are closed, sometimes obliterated in adult specimens.	19. Sutures of the skull and of epiphyses of the skeleton are distinct in fully adult individuals (Sinclair). With this the writer's observation agrees.
20. Humerus is without trace of an entepicondylar foramen.	20. A large entepicondylar foramen is present in *Prothylacynus* and in *Cladosictis*. *Amphiproviverra* is without such a foramen (Ameghino).
21. Basicranial axis forms a strong angle with the basifacial region.	21. Basicranial and basifacial planes are parallel.
22. Posterior root of the zygoma descends below the basicranial axis.	22. Posterior root of the zygoma is in the plane of the basicranial axis.
23. No transverse canal.	23. No transverse canal. This disagrees with most marsupials outside the Borhyænidæ.
24. The alisphenoid has no contact with the parietal.	24. Doubtful, specimens do not show.
25. Forefoot is digitigrade, the hind foot plantigrade.	25. The position of the feet is more or less conjectural. They were both figured as plantigrade by Sinclair.
26. The lunar is small, the magnum relatively large. The two have contact.	26. The same is true of *Borhyæna*.

CONCLUSION

Thylacosmilus is a new and a very distinct genus of marsupial carnivore. It is related to the Borhyænidæ of South America. This group has a known range from Eocene to the close of the Pliocene but had its greatest known dispersal in the Santa Cruz or upper Miocene. *Thylacosmilus* is known from two species which occur in the Araucanense deposits of Pliocene age, Catamarca, northern Argentina. This genus is an aberrant and highly specialized member of the Borhyænidæ, attaining a size considerably greater than that of the well-known *Borhyæna*. It is referred to a distinct subfamily, the Thylacosmilinæ.

The relations of *Thylacosmilus* to the other members of the family is that of a highly specialized branch to the more conservative members. The distinctions are based chiefly upon the peculiar structure of the facial region of the head, the mandible and to some extent the neck. The outstanding feature is the development of the upper canine teeth into saber teeth and the resultant modifications of the skull and the mandible to conform to their growth and use.

There are no known ancestors of *Thylacosmilus* among the South American borhyænids. It is very similar in heritage characters to such borhyænids as *Borhyæna* and *Prothylacynus* of the Santa Cruz. Species of *?Acrohyænodon*, contemporary with it, are known only from fragments of teeth and mandibles, but apparently are widely removed from it in structural characteristics.

In its specialized structure, *Thylacosmilus* shows a remarkable parallelism with the sabertooth tigers. The development of the canine tooth took on a similar form in the two widely separated groups of animals. It took root more deeply in *Thylacosmilus* and so caused a much greater modification of the facial region. It overshadowed and crowded out the incisors above and below; it produced a similar but less extensive reduction in the premolar series; and it developed a much deeper mandibular symphysis and a much larger protecting flange. The habits of the animals and the method of killing their prey may well have been similar.

The disappearance of *Thylacosmilus* from the fossil-bearing formations of South America took place soon after the appearance of placental carnivores on that continent. The procyonids of large species are known to have been contemporary with members of this genus in the Araucanense of Catamarca. Ursids reported by Ameghino from the Entrerriano of Paraná have been later questioned, but are known to have become widely distributed in Pleistocene stages; canids are earliest known from the base of the Pampean, machairodonts are common in that and other Pleistocene formations. It is quite reasonable to infer that the sharper competition introduced with the appearance of these placental carnivores was responsible for the elimination of the marsupial sabertooth which in its turn had been the most highly specialized, the strongest and no doubt the most destructive of all the long line of South American marsupial carnivores.

AUTHORS CITED IN THIS PUBLICATION

AMEGHINO, F.
    1886. Contribuciones al Conocimiento de los Mamíferos Fósiles de los Terrenos Terciarios del Paraná. Bol. Acad. Nac. Cienc. Córdoba, IX, pp. 1–228.
    1898. Sinopsis geológico-paleontólogica. Segundo Censo de la República Argentina, B.A., Tomo I, Buenos Aires.
    1903–4. Neuvas Especies de Mamíferos, Cretáceos y Terciarios de la República Argentina. An. Soc. Cient. Arg., LVI, pp. 193–208.
BENSLEY, B. A.
    1903. On the Evolution of the Australian Marsupialia; with Remarks on the Relationships of the Marsupials in General. Trans. Linn. Soc. London, Ser. II, Vol. IX, pp. 83–217.
BRAVARD, A.
    1858. Mongrafia de los Terrenos Terciarios de los Cercanos del Paraná. Paraná, 1858 (not seen).
BURMEISTER, H.
    1885. Examen Crítico de los Mamíferos y Reptiles Fósiles Denominados por Auguste Bravard. An. Mus. Nac. Hist., Buenos Aires, Ent. XIV.
    1892. Adiciones al Examen Crítico de los Mamíferos Fósiles Tratados en al Articulo IV Anterior. Ibid., III, pp. 375–400.
CABRERA, A.
    1927. Los Dasiuroideos Fósiles Argentinos. Rev. del Mus. de la Plata, Vol. XXX, pp. 271–315.
GREGORY, W. K.
    1910. The Orders of Mammals. Bull. Am. Mus. Nat. Hist., Vol. XXVII, pp. 1–524.

Riggs, E. S.
    1933. Preliminary Description of a New Marsupial Saber-tooth from the Pliocene of Argentina. Geol. Ser. Field Mus. Nat.
        Hist., Vol. VI, pp. 61–66.
Rovereto, C.
    1914. Los Estratos Araucanos y sus Fósiles. An. Mus. Nac. Hist., Buenos Aires, Tomo XXV, pp. 1–250.
Sinclair, W. J.
    1905. The Marsupial Fauna of the Santa Cruz Beds. Proc. Am. Phil. Soc., XLIV, pp. 73–81.
    1906. Mammalia of the Santa Cruz Beds, Marsupialia. Reports of the Princeton University Expeditions to Patagonia,
        Vol. IV, pt. 3, pp. 333–460.
Simpson, G. G.
    1932. Skulls and Brains of Some Mammals from the Notostylops Beds of Patagonia. Am. Mus. Novitates, No. 578.
Wood, H. E.
    1924. The Position of the Sparassodonta. Bull. Am. Mus. Nat. Hist., Vol. LI, pp. 77–101.

APPENDIX

Synonymy of Eutemnodus americanus Burmeister

By Bryan Patterson

Eutemnodus americanus Bravard, 1858 (nomen nudum). Monog. Terrenos Terciarios de los Cercanos del Paraná. Paraná,
    1858, 16.
Eutemnodus americanus P. Gervais, 1867–9 (nomen nudum). Zool. et Paleont. gén., I, 130.
Eutemnodus americanus H. Gervais and Ameghino, 1880 (nomen nudum). Les mamifères fossiles de l'Amérique du Sud, Paris,
    1880. (See also p. 527 in Obras Completas de F. Ameghino, Vol. 2.)
Eutemnodus americanus Burm., 1885. An. Mus. Nac. B.A., Ent. 14, 97–98, Pl. 3, fig. 1.
Hyænodon Lydekker, 1885. Cat. Foss. Mamm. Brit. Mus., I, 21 (footnote).
Apera sanguinaria Ameghino, 1886. Bol. Acad. Nac. Cienc. Córdoba, 9, 13–14.
Apera sanguinaria Ameghino, 1889. Act. Acad. Nac. Cienc. Córdoba, 6, 340–341, 913–914, Pl. i, Figs. 27–28. Pl. 77, Figs. 1–3.
Hyænodon sudamericanus Burm., 1891. An. Mus. Nac. B.A., 3, 375–376.
"Eutemnodus americanus" (Apera sanguinaria) Amegh., 1891. Rev. Arg. Hist. Nat., 260–261.
Apera sanguinaria Amegh., 1892. Bol. Acad. Nac. Cienc. Córdoba, 12, 465.
Eutemnodus americanus Trouessart, 1898, Catalogus Mammalium etc., New edition, pt. 5, 1215.
Eutemnodus americanus Palmer, 1904. Index generum mammalium, 282, 888.
Apera sanguinaria Palmer, 1904. Ibid. 111, 888.
Hyænodon americanus Amegh., 1904. An. Soc. Cient., Arg. 58, 266.
Hyænodon americanus Amegh., 1906. An. Mus. Nac. B.A. (3), 8, 392, 393–394, Fig. 255.
Apera americana Simpson, 1928. Post-Mesozoic Marsupialia. Fossilium Catalogus, pars 47, 43.

Type: A lower molar, M_1 according to Burmeister, the last molar according to Ameghino.

Bravard, P. Gervais, H. Gervais and Ameghino gave no description of this form. Burmeister (1885) described some of Bravard's fossils from Paraná. Among them was a crown of a molar which he held was the type of *Eutemnodus americanus*. The name must therefore be credited to him. He considered that the specimen was the lower carnassial of an extinct cat distinguished from *Felis* by the lacelike pattern of the pits on the surface of the enamel. At the same time he described an incisor which also had the characteristic pitting.

Lydekker, in the same year (1885), stated that the name *Eutemnodus*, as used in Bravard's MS., was a synonym of *Hyænodon*. The British Museum had acquired the Bravard collection * and in the catalogue of that collection Bravard had employed the name to cover certain species which Lydekker considered as belonging to *Hyænodon*. It should be noted that the name *Eutemnodus* was never published in connection with a European fossil, but only in connection with the South American *E. americanus*.

Ameghino (1886) suspected that an error had been committed somewhere and inclined

---

* Apparently only the European collection.

to the belief that Burmeister had been mistaken in referring the animal described by Bravard to *Eutemnodus*. This last, he states, is an extinct European genus founded by Pomel and Bravard and considered by them to be near *Didelphis*.* Therefore, since he accepts Burmeister's reference of *E. americanus* to the cats, he states that there can be no *Eutemnodus americanus* and renames, as *Apera sanguinaria*, the specimen which Burmeister had figured and described. Ameghino's creation of a new specific name was entirely unjustified.

In the main body of his great work on South American fossils Ameghino (1889, pp. 340, 341) expresses doubt about the correctness of Burmeister's reference of the animal to the cats but states that, not having examined the actual specimens, he is unable to confirm his suspicions. He also suspects that Burmeister was incorrect in referring the tooth he described to *Eutemnodus*. He argues that, since Lydekker has shown that *Eutemnodus* is a synonym of *Hyænodon*, Bravard, who was thoroughly familiar with *Hyænodon*, could hardly have applied the name *Eutemnodus* to an animal so distinct from *Hyænodon* as *Apera sanguinaria* (= *Eutemnodus americanus* of Burmeister). Subsequent to writing this, however, he obtained several other teeth (P^1, $P_{\overline{3}}$ and $P_{\overline{4}}$) which he referred to his *Apera sanguinaria* on the grounds that they show the characteristic pittings of the tooth described by Burmeister. From an examination of these teeth he became convinced that he was dealing with a creodont similar to *Hyænodon*, and came to agree with Bravard that a hyænodont had lived in the Tertiary of South America (Ameghino 1889, pp. 913–914, appendix to the main body of the work).

Burmeister (1891) mentions Lydekker's work which had appeared about the same time as his own, viz. 1885. He states that he has obtained another tooth, the penultimate upper, which agrees very closely with that of *Hyænodon leptorhynchus*. He therefore feels that *Eutemnodus americanus* should be referred to *Hyænodon* but probably not to *H. leptorhynchus*. He proposes a new specific name "*sudamericanus*" which he considers more appropriate as there are already several species of the genus known in North America.

Ameghino (1891 and 1892) criticises Burmeister for changing the specific name but refuses to admit that it is a *Hyænodon*, retaining his own name of *Apera sanguinaria*. In 1904, however, he reversed his position and declared that the real name was *Hyænodon americana* (Bravard) and gave the complete synonymy. In 1906 he gave figures of the teeth and some further description.

The fact which appears to have misled Burmeister and Ameghino was that the name *Eutemnodus* was, so far as I can ascertain, never applied to a European fossil except in MS., and that Bravard's nomen nudum was the first published mention of it. This fact is brought out in the catalogues of Trouessart and Palmer. Both of these writers, however, overlooked the establishment of the name by Burmeister in 1885. The teeth of the animal have a resemblance to those of *Hyænodon* but it is extremely improbable that any creodonts ever lived in South America. *Eutemnodus americanus* therefore stands as the name of a peculiar carnivorous marsupial of the Entrerrios beds.

FIELD MUSEUM OF NATURAL HISTORY, CHICAGO.

* Ameghino was mistaken in stating that *Eutemnodus* was described by Pomel and Bravard as being near *Didelphis*. *Hyænodon* was described by Laizer and Parieu in 1838 as a subgenus of *Didelphis*.

PLATE I

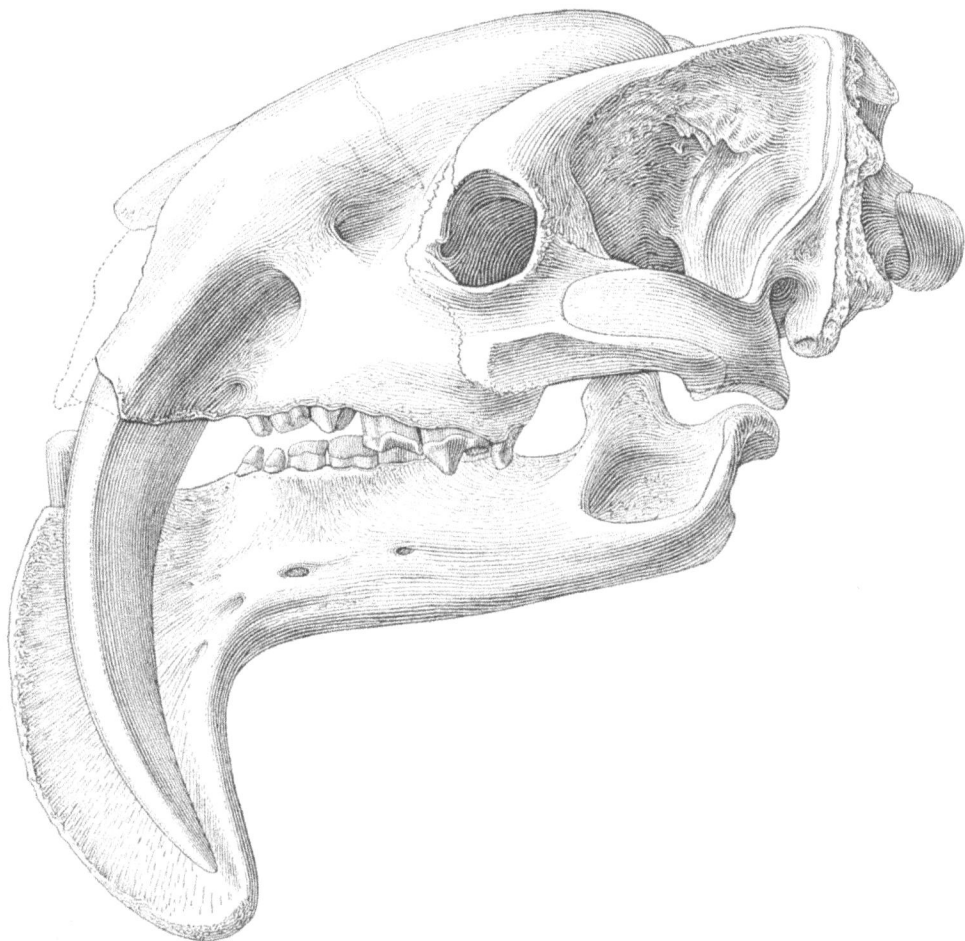

Thylacosmilus atrox. Holotype, P14531. × ⅗. Lateral view of skull. Mandible restored according to that of the paratype. Drawing by Sydney Prentice.

PLATE II

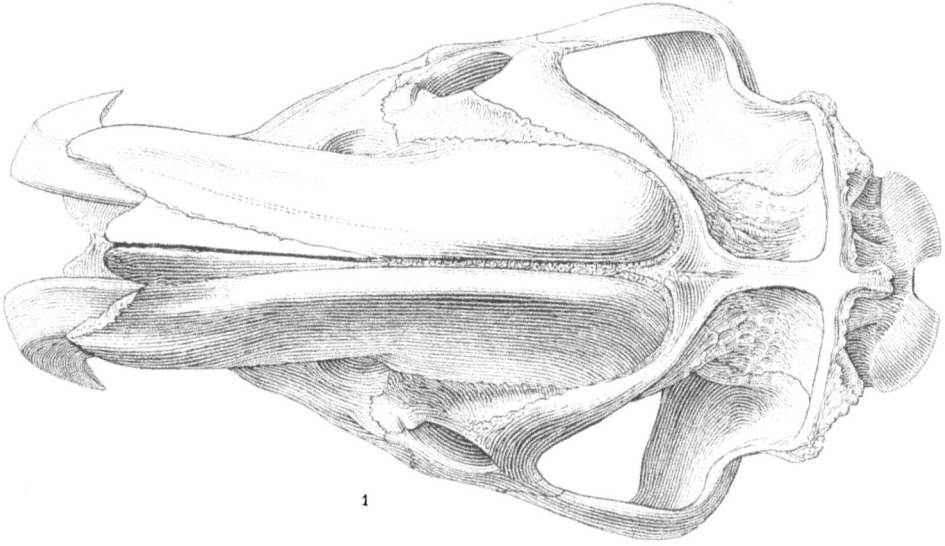

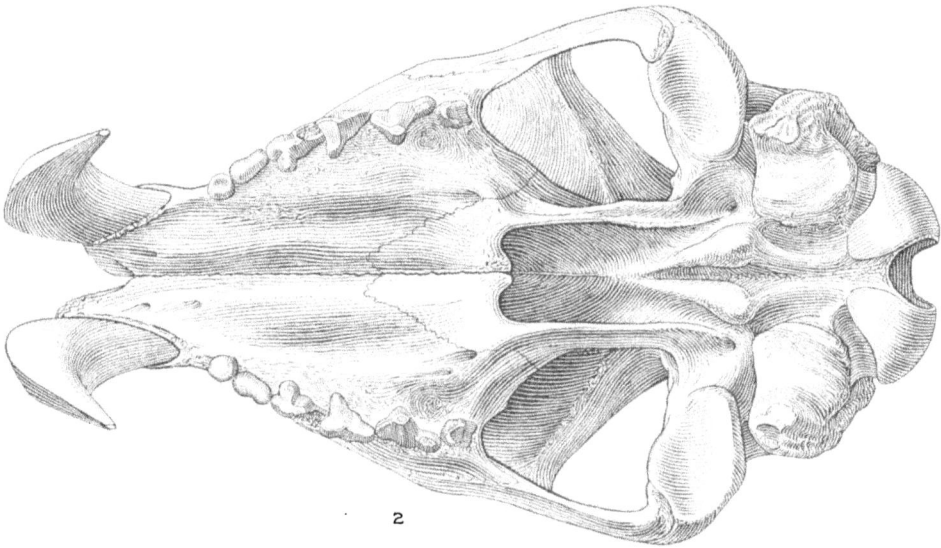

Thylacosmilus atrox. Holotype, P14531. × ⅗. 1, Dorsal view of the skull. 2, Palatal view of the same.
Drawings by Sydney Prentice.

PLATE III

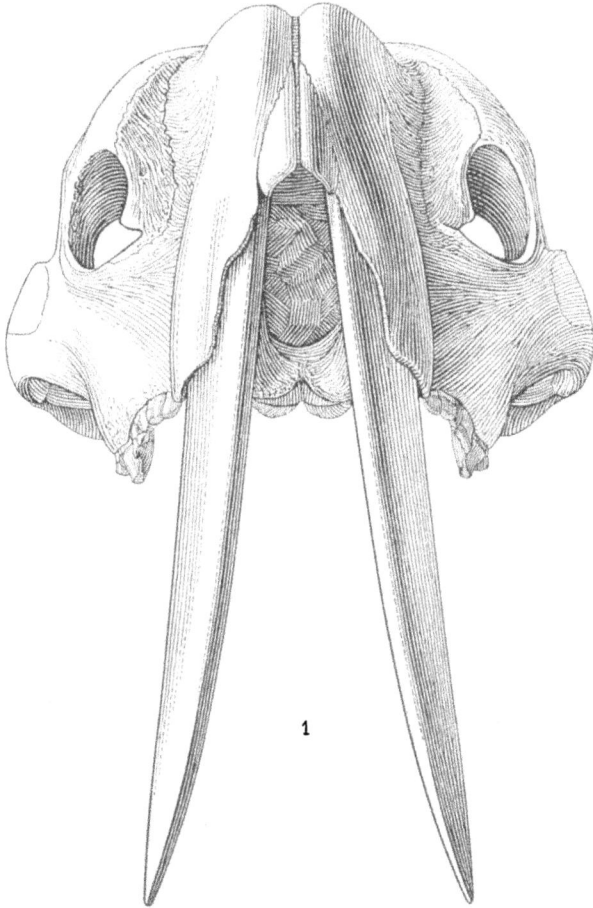

1

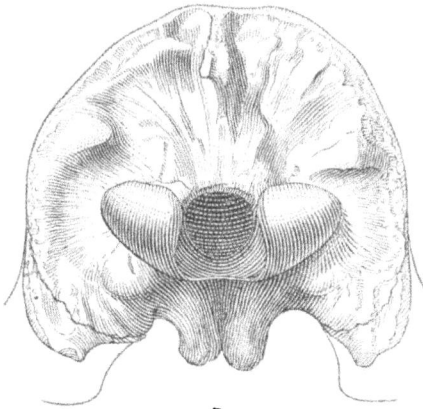

2

Thylacosmilus atrox. Holotype, P14531. × ⅔. 1, Anterior view of the skull. 2, Posterior view of the same.
Drawings by Sydney Prentice and Carl F. Gronemann.

PLATE IV

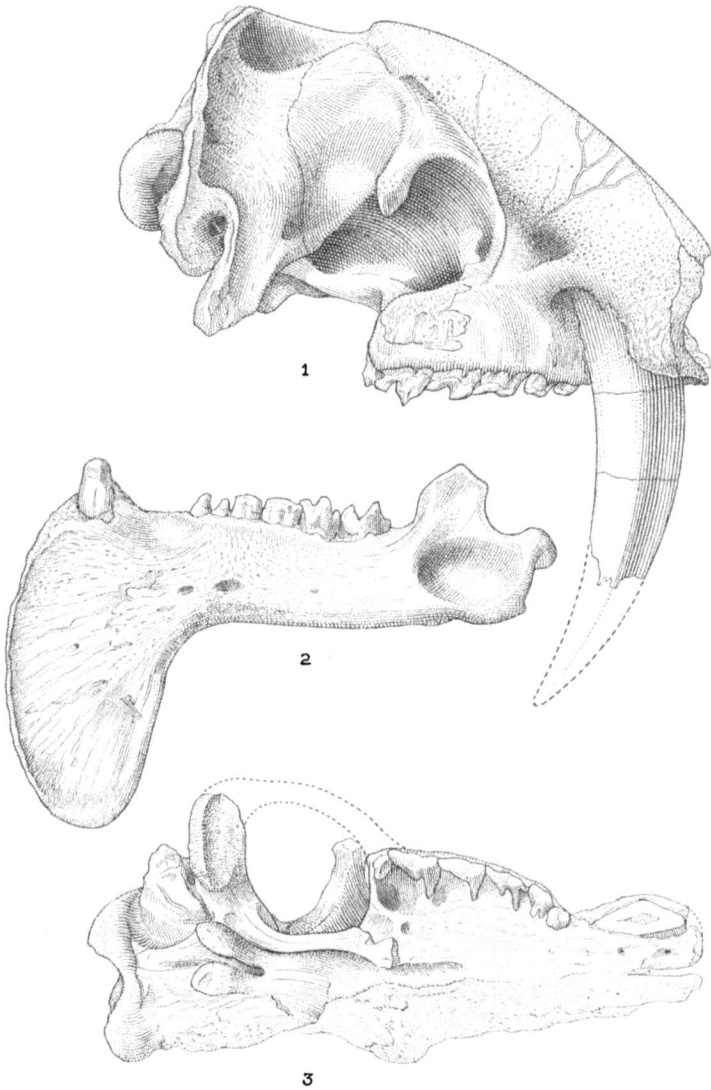

Thylacosmilus lentis. Holotype, P14474. × ⅔. 1, Lateral view of the skull. 3, Palatal view of the same.
T. atrox. Paratype, P14344. × ⅔. 2, Lateral view of the mandible.
Drawings by Carl F. Gronemann.

PLATE V

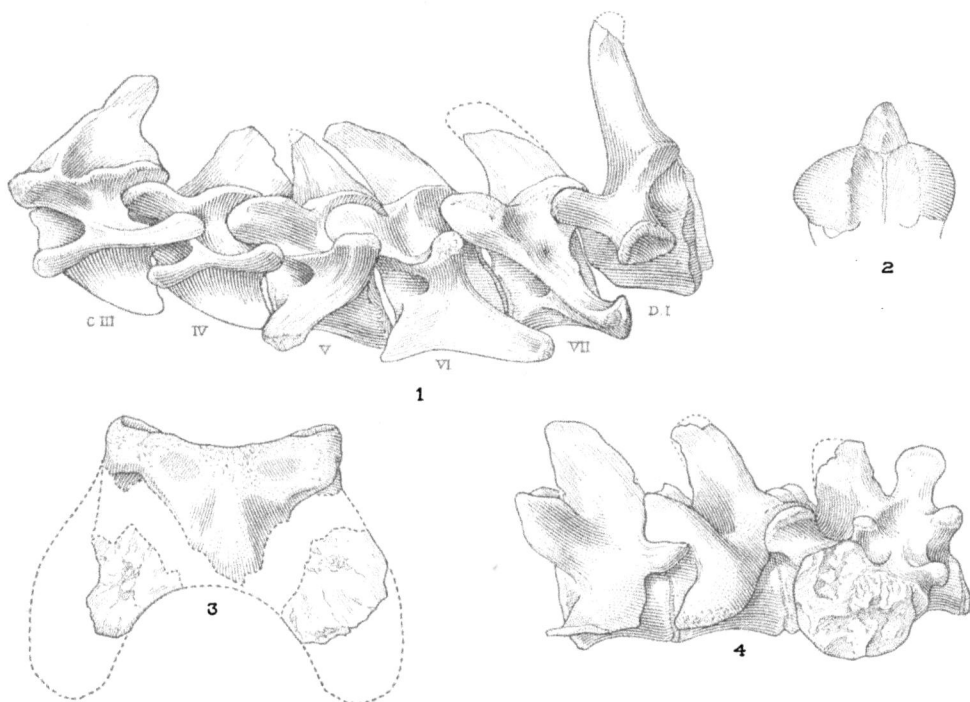

Thylacosmilus atrox. Paratype, P14344. × ⅔. 1, C. III–VII, lateral view of cervical vertebræ; D. I, lateral view of first dorsal vertebra. 4, Lateral view of two posterior lumbar vertebræ and of sacrum in series. *T. atrox.* Holotype, P14531. × ⅔. 2, Anterior end of the axis. 3, Dorsal view of the atlas.

Drawings by Carl F. Gronemann.

PLATE VI

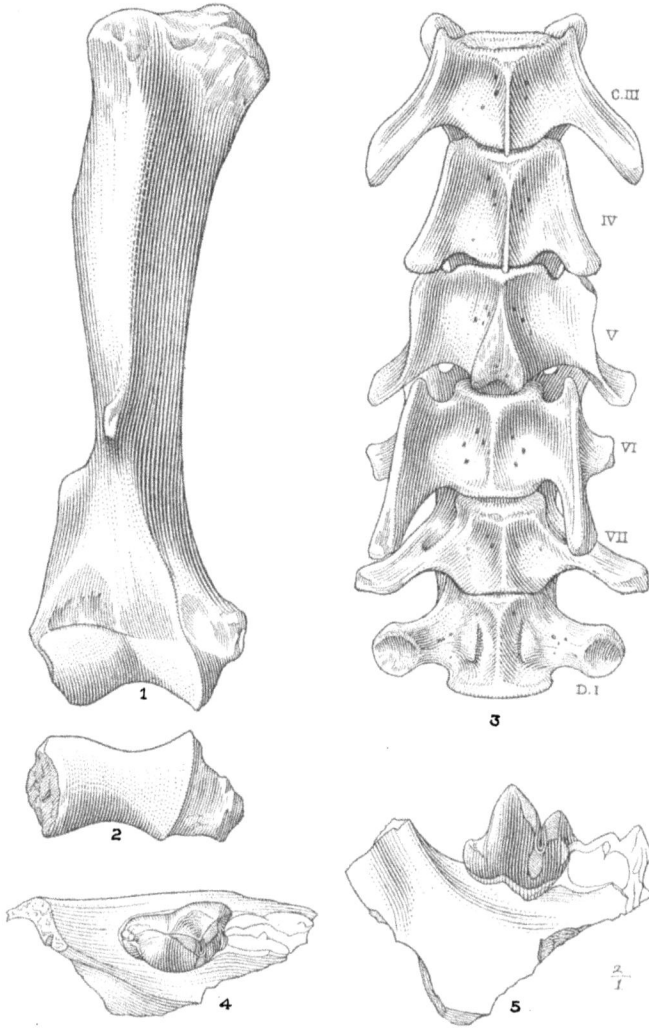

Thylacosmilus atrox. Holotype, P14531. × ⅔. 1, Right humerus, anterior view. 2, The same, distal end. Paratype, P14344. × ⅔. 3, C. III–VII, ventral view of cervical vertebræ; D. I, anterior dorsal vertebra. *Acrohyænodon acutidens.* P14407. × ⁴/₃. 5, Fragment of mandible with fourth molar, crown view. 5, The same, lateral view.

Drawings by Carl F. Gronemann.

PLATE VII

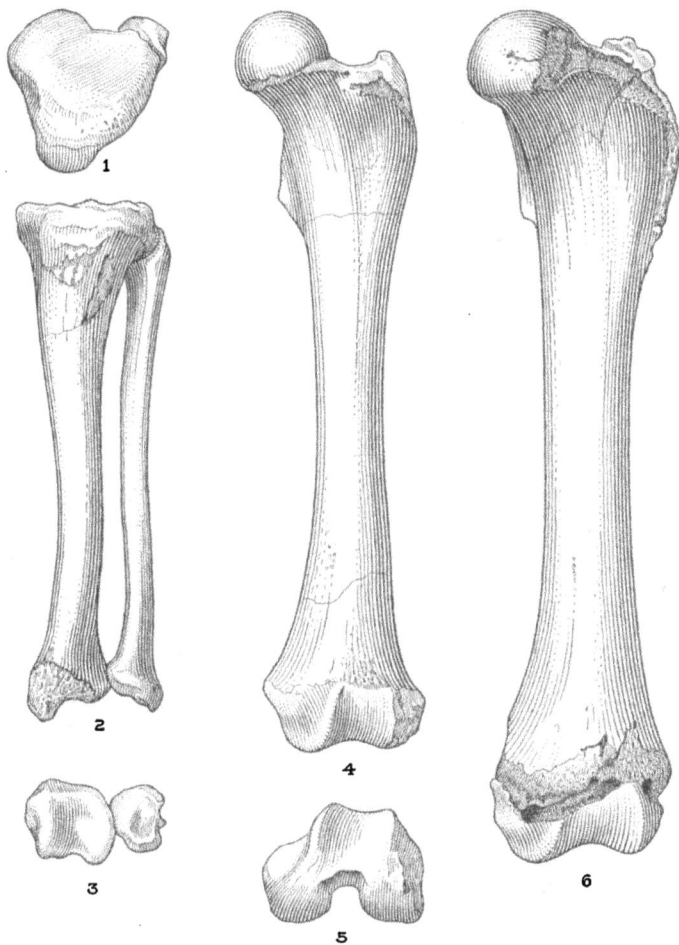

Thylacosmilus atrox. Paratype, P14344. × ⅔. 1, Proximal end of the left tibia. 2, Anterior view of the left tibia and fibula. 3, Distal ends of the same. 4, Anterior view of the left femur. 5, Distal end of the same. *T. atrox.* Holotype, P14531. × ⅔. 6, Anterior view of the left femur.

Drawings by Carl F. Gronemann.

PLATE VIII

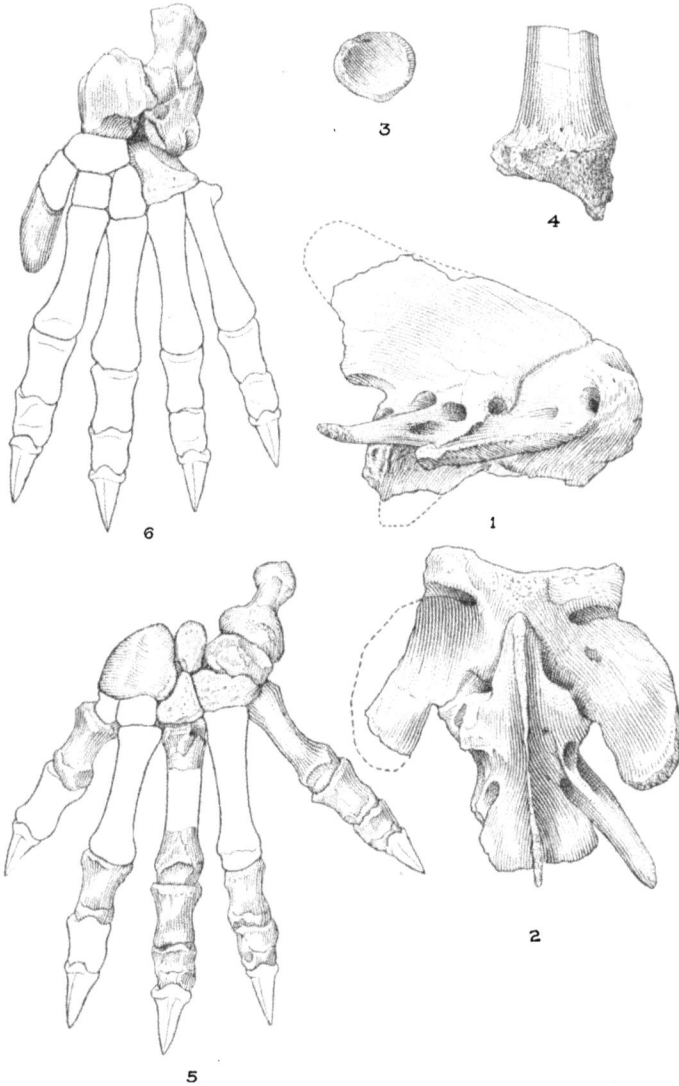

Thylacosmilus atrox. Paratype, P14344. × ⅔. 1, Lateral view of the atlas and axis. 2, Dorsal view of the same. *T. atrox.* Holotype, P14531. × ⅔. 3, Proximal end of the right radius. 4, Distal end of the same. 5, Left fore foot with scaphoid and fifth metacarpal restored after those of the right foot. 6, Left hind foot, of which the astragalus, calcaneum, cuboid and vestigial met. ı are known. Parts in outline are restored by comparison with the foot of *Thylacynus.*

Drawings by Carl F. Gronemann.

www.ingramcontent.com/pod-product-compliance
Lightning Source LLC
Chambersburg PA
CBHW070243230326
41458CB00100B/5974